LOGICALLY DETERMINED DESIGN

LOGICALLY DETERMINED DESIGN

CLOCKLESS SYSTEM DESIGN WITH NULL CONVENTION LOGIC™

Karl M. Fant
Theseus Research, Inc.

WILEY-INTERSCIENCE

A JOHN WILEY & SONS, INC., PUBLICATION

Published by John Wiley & Sons, Inc., Hoboken, New Jersey.
Published simultaneously in Canada.

For general information on our other products and services please contact our Customer Care Department within the U.S. at 877-762-2974, outside the U.S. at 317-572-3993 or fax 317-572-4002.

Wiley also publishes its books in a variety of electronic formats. Some content that appears in print, however, may not be available in electronic format.

Library of Congress Cataloging-in-Publication Data:
Fant, Karl M.
  Logically determined design: clockless system design with NULL convention logic™/Karl M. Fant.
    p. cm.
  "A Wiley-Interscience publication."
  Includes bibliographical references and index.
  ISBN 0-471-68478-3
  1. Asynchronous circuits–Design and construction. 2. Logic, Symbolic and mathematical.
  3. Logic design. 4. Computer architecture. I. Title.

TK7868.A79F36 2005
621.381–dc22   2004050923

Printed in the United States of America

10 9 8 7 6 5 4 3 2 1

To Michelle, Lara, and Omi

CONTENTS

Complexity is a primary problem of contemporary digital system design. While manufacturing productivity is growing at 60% a year, design productivity is growing at 25% a year [41]. The industry is mobilizing with more complex tools and more complexly integrated tools. The problem of complexity is being conquered with complexity.

The possibility has not been considered that the practical complexity being encountered might be considerably greater than it needs to be. Where the accepted conceptual foundations are inadequate to the task, and have unnecessarily compounded the inherent complexity of a system, the complexity of practice might be significantly diminished with a slightly different conceptual orientation.

As this book shows, a Boolean logic expression requires a carefully coordinated supplementary expression of a critical time relationship. This extra expression of time and its coordination complicates the expression of a digital system far beyond its intrinsic logical complexity. If there were a logic sufficiently expressive to not require supplementary expression, it would be possible to simply and straight-forwardly characterize a digital system and its behavior purely in terms of logical relationships, a logically determined system.

This book derives such a sufficiently expressive logic, NULL Convention Logic™, and presents the methodologies of designing logically determined systems. The book is an introduction to logically determined electronic system design for the interested scientist, engineer, or student. It presents conceptual foundations as well as basic system structures and practical methodologies of a logically determined electronic system design.

Chapter 1 explains how the accepted conceptual foundations of digital system design compound the inherent complexity of a system and presents a conceptual foundation to avoid this compounding of complexity. The practical embodiment of this conceptual foundation is developed in Chapters 2 through 4 in the form of 2 value NULL Convention Logic. Chapters 5 through 9 present the basic structures of a logically determined system design. Chapters 10 through 14 present the methodologies of understanding and managing the behavior of a logically determined system which is different from the familiar behavior of a clocked system.

A logically determined system exhibits what is referred to as asynchronous behavior. Although NCL encompasses many elements of traditional asynchronous design, this book is not about asynchronous design in its traditional context. The familiar elements are assembled here in a new context of a coherent and

xv

consistent logic. The interested reader can find the field of asynchronous design very ably presented in several contemporary texts.

For more information on this book please visit the following website: www.theseusresearch.com.

ACKNOWLEDGMENTS

Although NULL Convention Logic (NCL) is entirely my own intellectual initiative, I have accepted significant help in its development.

The concepts presented here were developed over many years with contributions from many people. Scott Brandt contributed to the discovery and conceptual development of NULL Convention Logic. Ken Wagner provided the business frame of mind as a partner, first in Theseus Research and then as a cofounder of Theseus Logic, Inc. in 1996. Theseus Logic is currently commercializing NULL Convention Logic. Over 20 chips have been successfully fabricated, and a commercially available tool set and macro library is available. Funding has come from private investment and from government research contracts with Ballistic Missile Defense Organization (BMDO), Defense Advanced Research Projects Agency (DARPA), National Security Agency (NSA), and the United States Army and Air Force. NCL is the property of Theseus Logic Inc. and a license must be obtained from Theseus logic for any commercial use of NCL.

Working with the engineers of Theseus Logic has provided me with a practical grounding and a deeper understanding of the technology. More specifically, Ryan Jorgenson provided the Excel Spreadsheet simulation of pipeline behavior used in Chapters 10 and 11, Mike Hagedorn invented the dual rail MUTEX of Chapter 8, and Steve Masteller developed the interweaved control pipelining of Chapter 8.

Michiel Ligthart, Ross Smith, and Oriol Roig formulated the first design methodology for NCL and produced the first commercially available tool set. Alexander Taubin and Alex Kondratyev joined Theseus Logic and contributed their deeply rooted academic background and expertise in asynchronous design to the development of the first design methodology, circuit techniques, 2D pipelining, and much more.

David Duncan absorbed the literature of the 1960s and 1970s on threshold logic design to understand the threshold nature of 2NCL. He defined the library of gates and several optimization techniques. He is primarily responsible for the content of Chapter 4. He also investigated the literature on pipeline behavior and contributed to Chapters 10 and 11.

Jerry Sobelman, Larry Kinney, and Keshab Parhi of the University of Minnesota brought early insights to the development of the technology. Jerry Sobelman formulated the gate implementations and with the assistance of Jason Hinze designed the early experimental test chips. Jordi Cortadella of the Universitat Politècnica de Catalunya contributed the synthesis experiment referred to in Section 4.7.2. Steve Furber of University of Manchester and Jens Sparso of the Technical University of Denmark provided valuable suggestions along the way.

Writing the book was supported by Theseus Logic.

CHAPTER 1

Trusting Logic

Boolean logic is a mathematical symbol system. There is a population of symbols organized into a state space, a set of primitive function mappings, a logic expression specifying a progression of primitive function mappings, and a set of rules of behavior to coordinate the flow of symbols from the state space through the progression of function mappings to resolve the logic expression. This passive symbol system is enlivened by an active agent that can manipulate the symbols in accordance with the rules, traditionally a mathematician with a pencil.

When a logic expression such as Figure 1.1 is enlivened by a mathematician, the logic expression and the mathematician form a complete expression that properly resolves. The mathematician, understanding the rules of behavior, coordinates the behavior of the symbolic expression with her pencil. She animates the symbols moving them from function to function, instantiating functions in the proper order when their input symbols are available. This coordination behavior is embodied in the mathematician's understanding of the rules of interpretation and is not explicit in the logic expression itself.

A function mapping simply specifies a mapping from input symbols to output symbols. Nothing is expressed about the presentation of symbols or about instantiation of the function. A logical expression specifies the dependency relationships among the input and output symbols of functions but expresses nothing about coordination behavior among the functions. On its own expressional merits, without the mathematician, a Boolean logic expression is an incomplete expression that cannot be trusted.

1.1 MATHEMATICIANLESS ENLIVENMENT OF LOGIC EXPRESSION

How the missing expressivity of the absent mathematician is restored to form a sufficiently complete expression is the crux of the matter. One can create a machine to emulate the mathematician. One can supplement the Boolean logic expression with

Logically Determined Design: Clockless System Design with NULL Convention Logic^{TM}, by Karl M. Fant
ISBN 0-471-68478-3   Copyright © 2005 John Wiley & Sons, Inc.

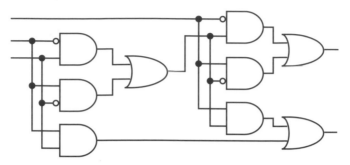

Figure 1.1   A Boolean logic expression.

some other form of expression. Or one can define a logic that is sufficiently expressive in its own terms to fully characterize reliable behavior.

1.2   EMULATING THE MATHEMATICIAN

One might suggest that the mathematician can be replaced by building the rules of coordination into an interpreting machine to properly resolve the logic expression. The logic expression remains a passive symbolic expression that is enlivened by the activity of the interpreter. This is a viable approach for all levels of abstraction except the most primitive. There must be an expression of a most primitive interpreter that cannot itself be interpreted. The logical expression of this last interpreter must behave spontaneously and autonomously on its own merits. The missing expressivity of the absent mathematician cannot ultimately be restored with an interpreting machine.

1.3   SUPPLEMENTING THE EXPRESSIVITY OF BOOLEAN LOGIC

Since the expression of the symbols, the function mappings, and the dependency relationships among functions express no boundaries of instantiation, mapping the logic expression to continuously acting spontaneous behaviors is a valid enlivenment of the symbolic expression. The functions can be represented such that they spontaneously and continuously transform symbols that spontaneously and continuously flow along the dependency relationships among functions. While these continuous behaviors faithfully enliven the logic expression itself, they do not encompass the coordinating expressivity of the missing mathematician.

1.3.1   The Expressional Insufficiency of Boolean Logic

When a new input is presented to a continuously behaving Boolean combinational expression, a stable wavefront of correct result transitions flows from the input through the network of logic functions to the output. Since Boolean functions are continuously responsive to freely flowing symbols and since some functions and signal paths are faster than others, a chaos of invalid and indeterminate result transitions may rush ahead of the stable wavefront of valid transitions. Such chaotic behavior causes the output of the expression to assert a large number of invalid results before the stable wavefront of correct results reaches the output and the expression as a whole stabilizes to the correct resolution of the presented input. A Boolean logic combinational expression cannot, on its own terms, avoid this indeterminate behavior.

But even if indeterminate transitions could be avoided, there is still no means to determine from the behavior of the expression itself when the output of the expression has stabilized to the correct result. If the current input state happens to be identical to the previous input state, the expression exhibits no transition behavior at all, correct or incorrect. Being unable to express its own boundaries of behavior a Boolean combinational expression cannot coordinate its own successive instantiations nor can it coordinate instantiations with other combinational expressions. These coordination behaviors must be expressed in some other way.

1.3.2 Supplementing the Logical Expression

A continuously behaving Boolean combinational expression can be relied on to eventually settle to a stable correct result of a stably presented input. After presentation of a new instance of input to a Boolean expression, it is necessary and sufficient to wait an appropriate time interval, characterized by the slowest propagation path in the expression, to ensure that the wavefront of stable results has propagated through the expression and the output of the expression has stabilized to the correct resolution of the presented input data. During the time interval all the erroneous transitions due to the racing wavefronts can be ignored, and at the end of the interval the correct result can be sampled. Thus the boundaries of instantiation and resolution can be expressed by supplementing the logic expression with an expression of a time interval. This time interval, however, is a nonlogical expression that must be associated with and coordinated with every Boolean combinational logic expression and that is specific to each implementation of a logic expression.

1.3.3 Coordinating Combinational Expressions

A combination logic expression will receive a succession of inputs to resolve and combinational logic expressions will pass results among themselves to resolve. For a given logic expression, each instance of resolution must be completed and stable before it can be passed on, and the result must be passed on before the given logic expression can accept a new input to resolve. These boundaries of behavior are expressed by the time interval associated with each combinational expression, so the coordination of behavior among combinational expressions must be expressed in terms of these associated time intervals. This coordination is accomplished with a state-holding element between each combinational logic expression that ignores transition behavior during an interval, samples the output of each logic expression at the end of each interval, and stably presents the sampled output as input to a successor combinational expression.

A system consists of multiple combinational expressions, each with an associated time interval and coordinated among themselves in terms of these time intervals. This coordination is most conveniently expressed when all the time intervals are the same duration and are in phase. All logic expressions present valid output and become ready for new input simultaneously, in phase with a common interval expression beating the cadence of successive instantiations.

While the coordinating expressivity of the mathematician has not been restored at the function level, it is sufficiently restored at the level of the time interval to enable practical application.

1.3.4   The Complexity Burden of the Time Interval

Supplementing Boolean logic with an expression of time compounds the complexity of the expression. The structure of the logical expression must be specified. The structure of the time expression must be specified. The delay behavior of the logic in a particular implementation must be determined. The behavior of the time expression in the particular implementation must be determined. The logic expression and the time expression must be integrated. The behavior of the combined expressions must be verified.

There can be no first-order confidence in the behavior of the combined expression. Confidence can only be gained in terms of exhaustive observation of its behavior. At the granularity of the time interval the combined expression generates a sequence of stable states that can be sampled and compared to a theoretically derived enumeration of states to verify correct behavior.

1.3.5   Forms of Supplementation Other Than the Time Interval

There have been ongoing attempts to provide supplemental forms of expression for continuously acting Boolean logic expressions other than the time interval called 'asynchronous design research.' The expressional insufficiency of Boolean logic was recognized by Muller in 1962 [35] who introduced two supplementary forms of expression.

The first supplement, called a C-element, could enhance Boolean expressions to avoid indeterminate output. A C-element is a state-holding operator that transitions its output to 1 only when both inputs are 1 and transitions its output to 0 only when both inputs are 0, and for 01 and 10 inputs does not transition its output but holds its current state. The C-element has to be an indivisible primitive operator in its own right. It cannot be implemented in terms of Boolean logic functions. But the C-element, added to Boolean logic, creates a heterogeneous logic that is difficult to formalize, analyze, and synthesize and that still has subtle timing issues [31,1,8].

The second supplement was dual-rail encoding of binary data, in which binary symbols are represented with two wires: 01 represents FALSE and 10 represents TRUE; 00 represents the absence of data and 11 is illegal. Dual-rail encoding can express the absence of data and the presence of data, providing a logical expression of the boundaries of successive data presentations.

The C-element and dual-rail encoding remain standard elements of asynchronous design. The general approach has been to compose Boolean logic expressions enhanced with C-elements, timing assumptions (e.g., wires with zero delay or detailed timing coordination of specific signals within a logic expression) and dual-rail encoding. While this can be achieved to a certain degree, the heterogeneous nature of the expression results in arcane logic structures with subtle modes of behavior including critical timing relationships [4,33,58,59].

Some asynchronous design methodologies pursue the minimal expression of timing relationships. Muller's modules [35], Delay-Insensitive Minterm Synthesis [53], and Martin's methodology [29,30] have approached logically determined design but have not presented a coherent, easily understandable, and adoptable conceptual foundation.

Other methodologies pursue maximum speed and minimum size by minimizing the expression of logical relationships and simply embracing the increased critical timing relationships as an integral part of the design methodology [55,56,54].

The bundled data/micropipeline approach splits the difference. It uses a standard timed Boolean data path and an asynchronous control substrate that expresses the intervals as local clocks to the data path registers using a matched delay element associated with each combinational expression [57,15].

1.3.6 The Complexity Burden of Asynchronous Design

While the C-element and dual-rail encoding were steps in the right direction, they did not step far enough. Boolean logic was retained as a primary form of expression. The result was a heterogeneous expression methodology with subtle logic structures and complex behavior modes that neither enlightened understanding nor enabled practice.

The primary question of asynchronous design—How can Boolean logic be made to work without a clock?—is the wrong question. Boolean logic does not provide an adequate conceptual foundation and simply confuses the issues. The reliance on Boolean logic as a fundamental form of expression is a primary reason that asynchronous design is difficult, is distrusted, and has not been adopted.

1.3.7 The Cost of Supplementation

As logic expression becomes more complex, supplementary expression simply compounds the complexity of expression. Clearly, one should strive to avoid the necessity of supplementary expression.

1.4 DEFINING A SUFFICIENTLY EXPRESSIVE LOGIC

What if a sufficiently expressive logic were available, a logic that could completely express the behavior of a system solely in terms of logical relationships? There would be no need for any supplementary expression with its compounding complexity. Such a logic must symbolically express the boundaries of data presentation, and its operators must appreciate the expressed boundaries.

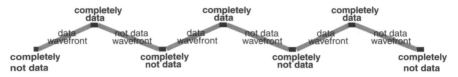

Figure 1.2 Monotonically alternating wavefronts of completely data and completely not data.

1.4.1 Logically Expressing Data Presentation Boundaries

First, the representation of the data must include an explicit symbolic expression of 'not data'. A symbol that explicitly means 'not data' is added to the 'data' symbols, and the input presented to the logic operators transition monotonically between 'completely not data' and 'completely data', as shown in Figure 1.2. The transition of the input from 'completely not data' to 'completely data', called a 'data' wavefront, expresses a new data presentation and the beginning of an instance of resolution. The transition of the input from 'completely data' to 'completely not data', called a 'not data' wavefront, expresses the end of an instance of resolution and the boundary between instantiations. Dual-rail encoding is a specific instance of such a representation.

1.4.2 Logically Recognizing Data Presentation Boundaries

To logically appreciate the presentation boundaries the logic operators must respond only to completeness relationships at their inputs in contrast to continuously responding to all possible inputs.

If input is:

- 'completely data', then transition output to the correct 'data' resolution of input,
- 'completely not data', then transition output to 'not data',
- neither 'completely data' nor 'completely not data' do not transition output.

Example operators are shown in Figure 1.3. T and F are the 'data' symbols and N is the 'not data' symbol. The dash means that the output of the operator does not transition.

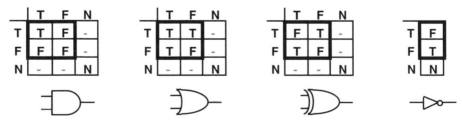

Figure 1.3 Logic operators that recognize 'data' from 'not data'.

Each logic operator, responding only to completeness of input relationships, coordinates its own behavior with the completeness behavior of the presented input. When the output of a logic operator monotonically transitions to 'completely data', it means that the input is 'completely data' and the data output is the correct result of the presented input. When the output of a logic operator monotonically transitions to 'completely not data', it means that the input is 'completely not data' and the 'not data' output is the correct result of the presented input. The C-element is a specific instance of such a logic operator.

This completeness behavior of the logic operator scales up for any acyclic combination of logic operators. Consider the combinational expression shown in Figure 1.4. Circles are logic operators and they are acyclically connected from input to output. Divide the network arbitrarily into N ranks of operators ordered progressively from input to output, with all inputs before the first rank and all outputs after the last rank. The rank boundaries are shown in Figure 1.4 with vertical lines labeled alphabetically in rank order from input to output. Consider that the expression is in a completely 'not data' state and a 'data' wavefront is presented. The 'data' wavefront will transition the inputs to 'data' as the input monotonically transitions to 'all data'.

- For the symbols crossing E to be all data all of the symbols crossing D must be data.
- For the symbols crossing D to be all data all of the symbols crossing C must be data.
- For the symbols crossing C to be all data all of the symbols crossing B must be data.
- For the symbols crossing B to be all data all of the symbols crossing A must be data.
- Therefore, for all the symbols after E to be data, all the symbols before A must be data.

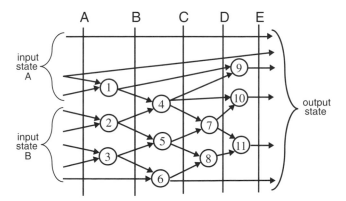

Figure 1.4 The completeness criterion for a combinational expression as a whole.

These considerations are also true for the 'not data' wavefront presented when the expression is in a 'completely data' state. Simply substitute 'not data' for 'data' in the text above.

The output of the combinational expression as a whole maintains the monotonic behavior of the input. When the output transitions to 'completely data', it means the input has transitioned to 'completely data' and the output is the correct resolution of the input. When the output transitions to 'completely not data', it means the input has transitioned to 'completely not data' and the output is the correct resolution of the input.

Given the monotonic behavior of the input, it does not matter when or in what order transitions occur at the input of the expression. Nor does it matter what the delays might be internal to the expression. Consider operator 7 in Figure 1.4. It does not matter how long the 'data' symbols ('not data' symbols) take to propagate through other operators and over signal paths to the input of operator 7, its output will not transition until all symbols are 'data' ('not data') at the input of the operator. For each wavefront, each logic operator synchronizes its input and switches its output exactly once coordinating the orderly propagations of monotonic transitions to correct result symbols through the combinational expression until the output of the expression as a whole is complete. There are no races, no hazards, and no spurious result symbols during the propagation of the monotonic wavefront of correct results through the combinational expression.

The behavior of the combinational expression is expressed entirely in terms of logical relationships. No expression of any time relationship or any other nonlogical relationship is necessary to fully characterize the behavior of the expression. The behavior of the combinational expression is completely logically determined.

1.4.3   Logically Coordinating the Flow of Data

Now that the boundaries of presentation and resolution can be expressed logically in terms of completeness relationships, the flow of presentation wavefronts among expressions can be coordinated in terms of these same completeness relationships.

When a combinational expression detects completeness on its output it generates an acknowledge signal in the inverse domain from the detected completion. For 'completely data' output the acknowledge signal transitions to 'not data'. For 'completely not data' output the acknowledge signal transitions to a 'data' symbol.

Each combinational expression includes an acknowledge signal as an integral part of its input completeness relation. When the acknowledge signal presented to a combinational expression transitions from 'not data' to 'data', the input of the combinational expression will become completely data and a data wavefront will be enabled to propagate to the output of the combinational expression. As long as the input acknowledge signal remains 'data' the combinational expression will stably maintain its 'data' output symbols even if a 'not data' wavefront is presented on the data path. When the acknowledge signal transitions from 'data' to 'not data', a 'not data' wavefront will be enabled to flow through the combinational expression. As long as the input acknowledge signal remains 'not data', a combinational

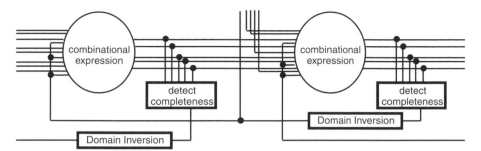

Figure 1.5 Coordination among combinational expression in terms of completeness relationships.

expression will stably maintain its 'not data' output symbols even if a 'data' wavefront is presented on the data path.

When a combinational expression detects completeness on its output, it means that a wavefront has been enabled through it by the acknowledge signal and that it is stably maintaining the output wavefront. Upon output completeness it can inform, via the acknowledge signal, each combinational expression whose output provided input to it that they need no longer maintain their output and that they can allow a new input wavefront to propagate. This relationship is illustrated in Figure 1.5. Stably maintained wavefronts flow through the system passed from combinational expression to combinational expression fully coordinated in terms of completeness relationships.

1.4.4 Mathematicianless Completeness of Expression

Symbols move from function to function, instantiating in the proper order as their input symbols are available. Wavefronts flow from expression to expression and instantiate in the proper order when their inputs are complete. This coordination behavior is explicit in the logic itself. The expressivity of the absent mathematician has been fully restored in the expressivity of the logic.

A logic expression can be complete and sufficient in itself. No supplementary expression is required.

1.5 THE LOGICALLY DETERMINED SYSTEM

A logically determined system is a structure of logical relationships whose behavior is completely and unambiguously determined by those logical relationships. Wavefronts spontaneously flow through combinational expressions fully coordinated by logical relationships. One might try to imagine the flowing wavefronts in terms of a single collective state, but this imagined collective state is subject to concurrent transitions coordinated solely by logical relationships with no consideration for any time relationships. The behavior of the collective state in relation to time is simply indeterminate. There is no way of sampling a stable configuration of the collective state that at any chosen instant might be in transition. The familiar methodology of understanding, and acquiring confidence by enumerating and comparing sequences of a collective state, is not applicable to a logically determined system.

1.6   TRUSTING THE LOGIC: A METHODOLOGY OF LOGICAL CONFIDENCE

Behavior that appears chaotic, undetermined, unknowable, and untrustable in the context of a collective state appears orderly, completely determined, directly knowable, and trustable in the context of logic relationships. The only path to understanding and trusting the behavior of a logically determined system is in terms of the logic relationships.

The behavior of a first logical expression can be understood locally in terms of its direct neighbor expressions. As the neighbor expressions are understood, their behavior neighborhood includes the first logical expression. As understanding progresses logical expression by logical expression, a tapestry of logically determined behavior is woven of overlapping behavior neighborhoods, Confidence in system behavior is progressively approached in humanly graspable steps.

1.7   SUMMARY

Boolean logic is insufficiently expressive on its own terms and must be supplemented with an expression of a mathematician or an expression of time or an expression of non-Boolean operators. The supplemental expression compounds the inherent complexity of the logical expression. The behavior of a supplemented expression cannot be directly trusted, so confidence in system behavior is found in exhaustive correspondence with enumerated states.

While the C-element and dual-rail encoding were steps toward sufficient expressivity, they did not step far enough. The C-element prefigures the importance of completeness behavior and dual-rail encoding prefigures the more general concept of monotonic transitioning between 'data' and 'not data'. The halter limiting the steps was the continuing reliance on Boolean logic as a primary form of expression.

Stepping far enough leads to a coherent logic that is sufficiently expressive to completely and unambiguously express the behavior of a system solely in terms of logical relationships. There is no need for any supplemental form of expression.

A system operating solely in terms of logical relationships behaves generally concurrently. There is no temporal synchronization of state transitions and, hence, no collective state that is reliably and meaningfully samplable. What appears to be chaotic and disorderly from the point of view of the state behavior is fully determined and orderly from the point of view of the logic. Confidence in system behavior requires a new rationale grounded in trusting the logic. A quite different regime of design, behavior, and rationale of confidence emerges. Since the behavior of a design is completely determined by the logical relationships, the logic can and must be trusted.

1.8 EXERCISES

1.1. Discuss the philosophical implications of a concept which is considered a primitive characterization but which must be supplemented to fulfill its mission of primitivity.

CHAPTER 2

A Sufficiently Expressive Logic

Another approach to overcoming the expressional insufficiencies of Boolean logic is to search for a practical logic that is expressionally sufficient. While the previous chapter defined the conceptual form of an expressionally sufficient logic, it did not define a practical instance of such a logic.

2.1 SEARCHING FOR A NEW LOGIC

The search is conducted by first enhancing Boolean logic to expressional sufficiency in terms of pure function operators. This results in a 4 value logic, which is impractical to implement. This 4 value logic is then evolved to a practical 2 value logic.

2.1.1 Expressing Discrete Data Presentation Boundaries

To the data values TRUE and FALSE is added the value NULL, to represent the state of 'no data'. This will be referred to as the NULL convention. The input presented to a combinational logic expression will monotonically transition between 'complete data' and 'no data' or all NULL. These monotonic transitions shown in Figure 2.1 will be referred to as wavefronts. The monotonic transition from 'complete NULL' to 'complete data' is a data wavefront and the transition from 'complete data' to 'complete NULL' is a NULL wavefront. Each successive data wavefront is separated by a NULL wavefront.

2.1.2 Logically Recognizing Discrete Data Presentation Boundaries

The disjoint domains of 'complete data' and 'complete NULL' can be appreciated by a logic function if a fourth value is added. INTERMEDIATE (I), which expresses input 'incomplete' (neither 'completely data' nor 'completely NULL'). The resulting 4 value logic functions are shown in Figure 2.2. The data function is shown in the bold box within the logic function table. A logic that includes a NULL value and that recognizes completeness relationships in the primitive logic operators will be

Logically Determined Design: Clockless System Design with NULL Convention Logic[TM], by Karl M. Fant
ISBN 0-471-68478-3 Copyright © 2005 John Wiley & Sons, Inc.

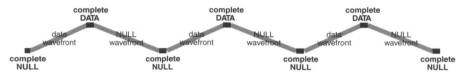

Figure 2.1   Successive wavefronts monotonically transitioning between disjoint domains.

referred to as a NULL Convention Logic (NCL) and this 4 value logic as 4NCL [10]. The recognition of completeness relationships will be called the completeness criterion.

Each function outputs a data value only when there is 'complete data' present at the input. Each function outputs a NULL value only when there is 'complete NULL' present at the input. Otherwise, the input is 'incomplete' and the function outputs an INTERMEDIATE value. If the input to a function monotonically transitions between 'complete data' and 'complete NULL' with possible INTERMEDIATE values during each transition, the function will maintain the monotonic transitioning behavior in its output.

2.1.3   The Universality of the NULL Function

4NCL logic functions vary in their behavior only for the data function. The NULL function behavior is identical for all operators and for any combination of operators. When presented with a NULL wavefront, they will all transition to NULL. So only the behavior of the data wavefront needs to be considered when specifying a combinational expression. The NULL wavefront behavior is universally identical and can be ignored.

2.1.4   Bounding the Behavior of a Combinational Expression

The completeness behavior of individual functions scales up for a combination expression as a whole. The discussion of a combinational expression behavior in Chapter 1 is reiterated here in terms of DATA, NULL, and INTERMEDIATE. Consider the combinational expression shown in Figure 2.3. The circles are logical

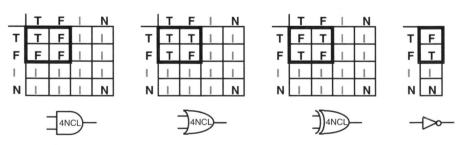

Figure 2.2   4NCL logic operator function tables.

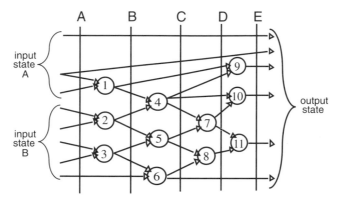

Figure 2.3   The completeness criterion for a combinational expression as a whole.

functions, and they are acyclically connected from input to output. Divide any such combinational expression arbitrarily into N ranks of functions ordered progressively from input to output, with all inputs before the first rank and all outputs after the last rank. The rank boundaries are shown with vertical lines labeled alphabetically in rank order from input to output.

For the values crossing E to be all data all of the values crossing D must be data.

For the values crossing D to be all data all of the values crossing C must be data.

For the values crossing C to be all data all of the values crossing B must be data.

For the values crossing B to be all data all of the values crossing A must be data.

Therefore, for all the values after E to be data, all the values before A must be data.

These considerations are also true for the NULL wavefront presented when the expression is in an all data state. Simply substitute NULL for data in the text above.

During a data wavefront value transitions progress monotonically from NULL to data through the expression from input to output. If any value crossing a boundary is NULL or INTERMEDIATE, then there will be at least one NULL or INTERMEDIATE value crossing all boundaries to the right of that boundary. There can only be a complete set of output data values when there is a complete set of input data values and the data values have propagated through the expression. There can only be a complete set of output NULL values when there is a complete set of input NULL values and the NULL values have propagated through the expression. The combinational expression as a whole expresses the completeness criterion and maintains the monotonic behavior of the input in its output.

The propagation of a wavefront through the expression is completely logically determined. Consider operator 7 in Figure 2.3. It does not matter how long the data values (NULL value) take to propagate through other functions and over signal paths to the input of function 7, or whether there are transitions through INTERMEDIATE, it will not output data (NULL) until all values are data

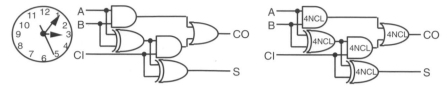

Figure 2.4 Boolean logic full adder and the corresponding 4NCL full adder.

(NULL) at the input of the function. No function transitions its output to data or NULL until all the values present at the input of the function have transitioned. Each function synchronizes the orderly propagation of monotonic transitions to correct result values through the combinational expression until the output of the expression as a whole is complete. There are no races, no hazards, and no spurious result values during the propagation of a wavefront through the combinational expression.

No expression of any time relationship, nor any other nonlogical relationship, is necessary to fully characterize the behavior of the expression. The behavior of the combinational expression is expressed entirely in terms of logical relationships and is completely determined by those logical relationships. It is a logically determined expression.

2.1.5 Relationship of 4NCL to Boolean Logic

Since 4NCL uses TRUE and FALSE to differentiate data values, 4NCL can be directly mapped function for function with a Boolean logic expression to form a logically determined 4NCL combinational expression as shown in Figure 2.4.

2.2 DERIVING A 3 VALUE LOGIC

A 4 value logic is not a practical logic for electronic implementation. The 4 value logic can be reduced to a 3 value logic that expresses the completeness criterion by replacing the expression of the INTERMEDIATE value with state-holding behavior as shown in Table 2.1. This reduction forms a 3 value NULL Convention Logic or 3NCL.

TABLE 2.1   Correspondences of 4NCL and 3NCL

4NCL	3NCL
TRUE	TRUE
FALSE	FALSE
INTERMEDIATE	State-holding behavior
NULL	NULL

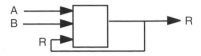

Figure 2.5   3NCL operator model with feedback path.

In 4NCL the INTERMEDIATE value explicitly indicates a partial completion state to be explicitly ignored. The partial completion states can also be ignored with a state-holding behavior. When the input of a logic operator is completely data, it outputs a data value and maintains the data value until its input is completely NULL, whereupon it outputs a NULL value and maintains the NULL value until its input is completely data, and so on. Instead of being explicitly expressed as a logic value by the logic operator, the partially complete states are simply ignored by the logic operator. Only wavefronts of NULL values and data values flow through the logic operators.

With state-holding behavior a logic operator is no longer a function and will henceforth be called an operator. Each operator still expresses the completeness criterion and 3NCL operators still compose to logically determined combinational expressions.

2.2.1   Expressing 3NCL State-holding Behavior

This state-holding behavior can be expressed as a function with feedback as in Figure 2.5 making the logic operator a state machine. The truth tables expressing the state-holding behavior for the 3NCL AND, OR, XOR, and NOT operators are shown in Figure 2.6. The NOT operator is a single input operator that cannot posses any intermediate input states, so it does not require state holding to explicitly express completeness for both data and NULL. The input R value is the output result

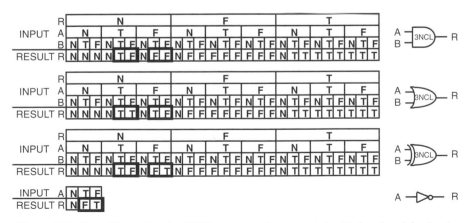

Figure 2.6   Transition tables for 3NCL operators implemented with functional feedback.

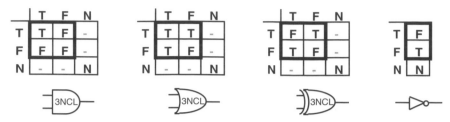

Figure 2.7   Transition table for 3NCL operators with natural state holding.

value fed back to the input. The data function of the operator is expressed by the values inside the bold boxes.

This feedback path must propagate and stabilize before the next wavefront is presented to the input of the operator. This timing relationship is noncritical and is easily satisfied.

If the operator is expressed in terms of a natural state-holding behavior such as the hysteresis behavior of magnetism, then there is no feedback path and no timing relationship issue. The implementation of the logic is completely logically determined. The tables in Figure 2.7 define the operator behavior for operators implemented with natural state holding behavior. A dash indicates that the output does not transition when the input is neither completely data nor completely NULL.

2.2.2 3NCL Summary

3NCL still differentiates two data values and a Boolean logic expression can be directly mapped into a logically determined 3NCL expression. 3NCL operators still express the completeness criterion and compose to logically determined combinational expressions.

With the functional feedback implementation of 3NCL operators, a timing issue arises with the feedback path. Clearly, this timing issue can be benignly isolated inside an operator and not compromise the logical determination of 3NCL expressions. Implementing the 3NCL operators with natural state-holding behavior presents no compromise whatever to the complete logical determination of 3NCL expressions. However, a 3 value logic is still impractical for electronic implementation.

2.3 DERIVING A 2 VALUE LOGIC

If the expressivity of the logic is reduced to only two logic values, it must be decided what those two values will differentiate. 4NCL and 3NCL differentiate data states from nondata states and also differentiate two data states. If only two logic values are available only one of these differentiations can be expressed.

If it is decided to differentiate two data states with the logical values such as TRUE and FALSE, the resulting logic is Boolean logic. If it is decided to

TABLE 2.2   Correspondences of the logics

4NCL	3NCL	2NCL	Boolean Logic
TRUE	TRUE		TRUE
FALSE	FALSE	DATA	FALSE
INTERMEDIATE	State-holding behavior	State-holding behavior	
NULL	NULL	NULL	

differentiate data from nondata, the NULL convention can be continued with the logical values DATA and NULL, and the state-holding behavior can be retained. The resulting logic is 2 value NULL Convetioin Logic(2NCL). Table 2.2 shows the correspondences between all four logics.

2.3.1   The Data Differentiation Convention

Since in 2NCL there is only one DATA value, each signal path in a 2NCL expression represents one specific meaning, and a path either asserts its meaning (DATA) or does not assert its meaning (NULL). To represent a variable that can express multiple mutually exclusive meanings (values), multiple signal paths must be used. To form a binary variable that mutually exclusively expresses the meanings TRUE and FALSE, for instance, there must be two signal paths, one signal path meaning TRUE and one signal path meaning FALSE. For any data wavefront only one of the two paths may express its DATA value. It is illegal for both DATA values to be simultaneously expressed. The two signal paths then form a single variable that can express two mutually exclusive DATA values or NULL.

In general, an N value NCL variable can be represented with N signal paths only one of which will assert a DATA value in a given wavefront. Figure 2.8 shows

Logical variable

	Signal meanings		NCL variable
	TRUE	FALSE	meanings
Signal	#1	#2	
Value	N	N	**NULL**
Value	**D**	N	**TRUE**
Value	N	**D**	**FALSE**

Numeric base 2 variable (binary)

	Signal meanings		NCL variable
	0	1	meanings
Signal	#1	#2	
Value	N	N	**NULL**
Value	**D**	N	**0**
Value	N	**D**	**1**

General three value variable

	Signal meanings			NCL variable
	Animal	Vegetable	Mineral	meanings
Signal	#1	#2	#3	
Value	N	N	N	**NULL**
Value	**D**	N	N	**Animal**
Value	N	**D**	N	**Vegetable**
Value	N	N	**D**	**Mineral**

Numeric base 4 variable (quaternary)

	Signal meanings				NCL variable
	0	1	2	3	meanings
Signal	#1	#2	#3	#4	
Value	N	N	N	N	**NULL**
Value	**D**	N	N	N	**0**
Value	N	**D**	N	N	**1**
Value	N	N	**D**	N	**2**
Value	N	N	N	**D**	**3**

Figure 2.8   Multi-path NCL variables.

several examples of NCL multi-value variables. The illegal states are not shown. This is a form of what has been called delay insensitive encoding. A 2 value variable is called dual-rail encoding [40,60].

Each signal path is a value of a variable. This is in contrast to the familiar situation of each signal path constituting a whole variable. Variables of any size can be built from signal paths and arbitrary mutually exclusive meanings can be assigned to the paths. This property is useful for implementing higher radix and even mixed radix numeric functions and for control expressions. For instance, a familiar one-hot control code, becomes a single large NCL variable. In this sense 2NCL is a general multi-value logic.

For a data wavefront, each variable of the wavefront asserts exactly one DATA value. Completeness of the data wavefront is exactly one DATA value per variable. For a NULL wavefront all asserted DATA values return to NULL and completeness for a NULL wavefront is all NULL values across all variables of the wavefront.

2.3.2   2NCL as a Threshold Logic

NULL means 'not data', so the NULL value cannot be considered in resolving data sets. Since each input to an operator expresses only one data value, there can be no combinations of different data values as in Boolean logic; there can only be combinations of a single value, DATA. The only discriminable property available when combining DATA values at the input to an operator is how many DATA values are presented. Therefore 2NCL is a threshold logic comprising discrete M of N threshold operators with state-holding behavior.

Figure 2.9 illustrates the graphical representation and the behavior of a 3 of 5 operator. The number inside the operator indicates the threshold of the operator M, and the input connections indicate the N. Wide lines are DATA. Thin lines are NULL. Beginning in a NULL state with all inputs NULL and asserting a NULL result value, the operator will not assert a DATA result value until its input data set reaches its threshold, which in this case is three DATA values. Then the operator will maintain the output DATA value until all of its input values are NULL at which time it will transition the output to a NULL result value.

Figure 2.10 shows the 2NCL family of logic operators with correspondences to the Boolean OR function and to the C-element. Weighted inputs are shown as

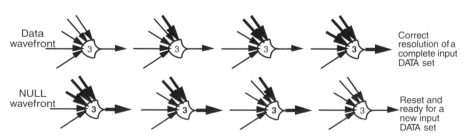

Figure 2.9   State-holding behavior of a 2NCL 3 of 5 operator.

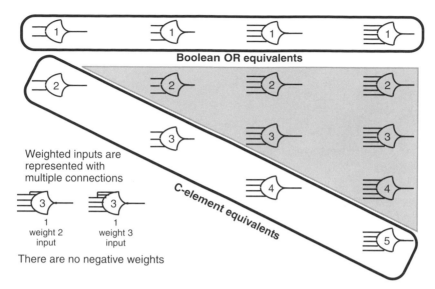

Figure 2.10 2NCL family of logic operators.

multiply connected inputs. There are no negatively weighted inputs. Threshold 1 operators (1 of N) do not require explicit state holding behavior. One DATA input will transition the output to DATA, and it will remain DATA until all the inputs are NULL. The behavior of the 1 of N operators is identical to the behavior of the Boolean OR function. The behavior of operators where M = N is identical to the C-element [35]. Notice that there is no equivalent to the Boolean AND operator and that there is no inverter. The operators in the gray field are unique to 2NCL.

Each operator and any combination of operators expresses the universal NULL function. The operators express a completeness criterion in relation to their thresholds. Their expression of the completeness criterion in relation to variables will be discussed below. While the logic lends itself to technologies that are inherently thresholding, the operators can be conveniently implemented in terms of CMOS switching circuits [50].

There is not as strong an intersection of 2NCL with classic threshold logic synthesis [27,36,42], as might be expected, because neither state-holding behavior nor completeness relationships were considered in the classic studies.

2.3.3 2NCL in Relation to Boolean Logic

While 2NCL is a complete and coherent logic in its own right and direct synthesis will be discussed in Chapter 4, it is instructive here to consider the relationships of 2NCL to the familiar Boolean logic. Figure 2.11 shows the 2NCL equivalents to the Boolean functions. Each Boolean function equivalent has two 2 value variable inputs and one 2 value variable output. Since the 2 value output variable comprises two signal paths there must at least two NCL operators to generate the two output signals.

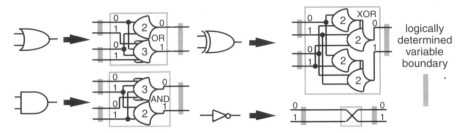

Figure 2.11   2NCL expression mappings for Boolean functions.

Notice that variable inversion does not involve signal path inversion. It is just a relabeling of the signal paths of the 2 value variable. Any single variable function for any size variable is just a mapping of each input value to an output value, and in the multipath representation this can be expressed by simply relabeling or rerouting the paths according to the mapping.

2.3.4   Subvariable Expressivity

With 2NCL the mapping to a Boolean expression is no longer operator for operator. A Boolean logic function is expressed as a combination of multiple 2NCL logic operators. While Boolean functions are dealing with inputs and outputs that are whole variables, 2NCL operators area dealing with inputs and outputs that are individual values of variables. An operation on variables is built out of operations on values individually. For instance, the 2 of 2 operator in the OR expression of Figure 2.11 has, as input, the 0 value path from each input variable but does not consider the 1 value path from either variable. In this sense 2NCL might be called a subvariable logic.

2.3.5   Completeness at the Variable Level

Logically determined completeness relationships are defined in terms of variables. The input and output boundaries of the 2NCL expressions of Figure 2.11 are variables. These expressions do not transition their output variables from NULL to DATA until the input variables are completely DATA (one DATA value per variable) and then do not transition their output variables to NULL until the input variables are completely NULL. The variable boundaries of each expression are logically determined completeness boundaries that express the completeness criterion for the expression as a whole. These 2NCL expressions can be directly substituted for Boolean functions in a Boolean combinational expression producing a logically determined 2NCL combinational expression.

2.3.6   The 2NCL Orphan Path

Internal to each 2NCL expression, the continuity of the variable is not maintained as each value path individually branches to many places. For each data wavefront through an expression there will be an effective path that is a logically determined irredundant path from the input through the expression that generates the output. When the output transitions to complete DATA, it implies that the input data set is complete and that the transitions to DATA have propagated over the effective path to the output. When the input transitions to NULL, the NULL values will propagate over the effective path to the output. When the output becomes all NULL, it means that the input is all NULL and that the effective path has transitioned to NULL.

There will also be ineffective paths branching off this effective path that do not contribute to the output and whose behavior therefore cannot be determined by the output behavior. These ineffective paths will be called *orphans* because they have lost all of their logical relations. Figure 2.12 shows the effective path and orphan paths for all four data configurations of the XOR expression.

A slow orphan path is not a problem for an individual data wavefront because the orphan paths do not contribute to the output, nor can they confuse the generation of the output. The difficulty arises with the possibility of a slow orphan not propagating NULL fast enough and getting a stale data value mixed up with a succeeding data wavefront.

Upon completion of the data wavefront at the output, the NULL wavefront will be requested and presented to the input of the expression while orphans may still be transitioning to DATA. The NULL wavefront will arrive, and the output of the expression transitioning to all NULL implies that the input and the entire effective path has transitioned to NULL; it does not imply that the orphans have transitioned to NULL. The output becomes completely NULL, and the next DATA wavefront is requested. If all the orphan paths from the previous data wavefront have not transitioned to NULL, the next data wavefront can interact with the slow orphan in a non-logically determined way. Ambiguous resolution, hazards, glitches, and chaos can ensue.

So the complete characterization of the behavior of a 2NCL expression must include the timing relationship that all orphan paths must completely transition before the succeeding wavefront arrives. In Figure 2.13 this orphan timing relationships is illustrated in the context of the coordination structure shown in Figure 1.5. In combinational expression Y the dark gray path is the orphan that must propagate from A to B before the light gray paths propagate from A through their respective

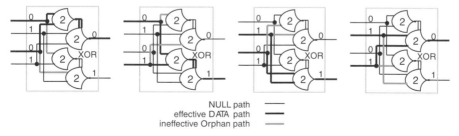

Figure 2.12   Effective paths and orphan paths through the XOR equivalent expression.

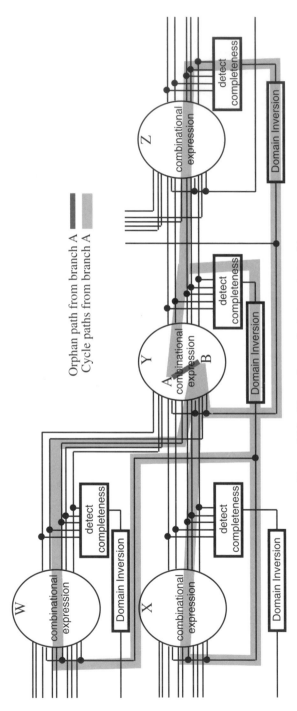

Figure 2.13 The orphan path-timing relationship.

combinational expressions, form completeness at the input of Y, and propagate to B. While the timing relationship of the orphan path must be kept in mind, it is a non-critical timing relationship that is easily managed.

Logical Isolation of Orphans. Orphan paths can easily be constrained to very local isolation within the logic structure. They do not cross logically determined variable completeness boundaries, and they can be isolated between these boundaries in any logical expression. If orphan paths crossed completeness boundaries, the boundaries would not be logically determined. Figure 2.14 provides an example of a 2NCL expression generated from a Boolean logic combinational expression by direct operator substitution. Each Boolean function equivalent expression is logically determined at its boundaries, and these boundaries become internal to the greater combinational expression. The effective path for $A = 0$ and $B = 1$ is shown with the associated orphan paths. It can be seen that all orphan paths are isolated between logically determined completeness boundaries of each Boolean operator equivalent expression.

It is possible for an orphan path to include 2NCL operators and to be arbitrarily long, but an orphan path can always be isolated to a local transmission path without an operator by inserting a logically determined variable boundary after the portion of the orphan path that includes operators and then subjecting the variable boundary to output completion determination. The part of the path including the operator then becomes an effective path that is logically determined by the output, thus shortening the orphan path to a local transmission path that does not include an operator.

Figure 2.15a shows a 2NCL expression with an orphan path that includes an operator. The outputs of the two 2 of 3 operators form an internal variable boundary that is not logically determined. Including the internal variable in the output completion in conjunction with output variables that are logically determined makes the paths of the orphan up to the point of branching to the output effective paths and hence makes them logically determinable. The remaining path of the orphan becomes a much reduced orphan path.

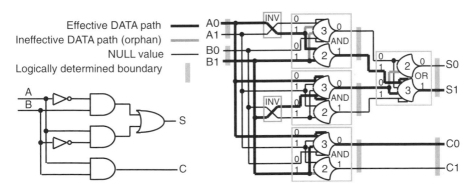

Figure 2.14 2NCL expression by direct substitution of a Boolean circuit.

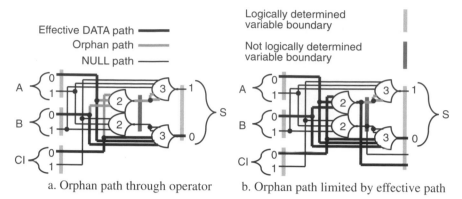

a. Orphan path through operator   b. Orphan path limited by effective path

Figure 2.15   Orphan path through operator isolated with a created output variable.

The orphan paths are the only logical ambiguity (delay sensitivity) associated with 2NCL. They can be conveniently isolated to local transmission paths inside an expression, and their delay behavior can be made insignificant in relation to the wavefront cycle period of a combinational expression. The timing sensitivity of the orphan is relative, not absolute.

2.3.7   2NCL Summary

2NCL is a practical, sufficiently expressive, 2 value logic suitable for electronic implementation. A logically determined electronic digital system can be completely expressed solely in terms of 2NCL. A noncritical timing relationship, the orphan, must be considered but can be easily managed.

Unlike 4NCL and 3NCL for which any combination of operators will be a logically determined expression, 2NCL operators can be combined into expressions that are not logically determined in any sense. So there must be rules of effective design associated with 2NCL. (The synthesis of 2NCL combinational expressions is discussed in Chapter 4.) Basically a 2NCL combinational expression must express the completeness criterion between input and output variable boundaries, and it must isolate orphan signal paths.

These rules can be isolated from the system designer by providing a library of 2NCL expression that are correctly formed and fully logically determined at their variable boundaries. The designer can then freely compose fully logically determined expressions simply by connecting the expression together through their variables. The first-order demonstration of this is the set of expressions of Figure 2.11, which are sufficient to map any Boolean combinational expression into a logically determined 2NCL combinational expression.

Logical Honesty. Logically determined variable boundaries are also integrity checking nodes as illustrated in Figure 2.16. The multi-path encoding is inherently

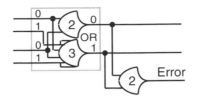

Figure 2.16   Error-detecting expression.

redundant and includes illegal codes. If two DATA values of any variable ever become 1 at the same time, this illegal condition can be detected by a 2 of N operator, with N being the number of values in the variable. For an 2NCL expression to lie, it requires two coordinated faults where a path that should remain NULL becomes DATA and the path that should become DATA remains NULL. Any stuck at fault will cause the monotonic transitioning to cease resulting in deadlock.

Each logically determined variable boundary will

1. assert a correct result, or
2. fail to assert a result (deadlock), or
3. assert a logically detectable illegal code.

2.4 COMPROMISING LOGICAL COMPLETENESS

With 2NCL one can choose to erode logical completeness for practical reasons. Eroding logical completeness means that less of the behavior of the expression is logically determined and more behavior must be expressed in terms of timing assumptions. To the extent that one can make reliable assumptions about expression behavior, one does not need to logically express that behavior. Of course, all the assumptions must then be made to be true in any actual implementation. In the case of 2NCL the consequence of eroding logical completeness is longer orphan paths with more critical timing relationships.

2.4.1 Moving Logically Determined Completeness Boundaries Farther Apart

One can choose, for instance, to compromise logical completeness by moving the logically determined variable boundaries farther apart. The 2NCL AND and OR functions can be expressed with simpler operators that do not express the completeness criterion. The expressions of Figure 2.17b expresses the same data transformation function as the expressions of Figure 2.17a, but they can generate an output without having a complete input data set. The output of the OR, for instance, can transition to 1 when only one input variable is data. The boundaries of the expressions are variables, but the variables do not express the completeness criterion and are not logically determined variable boundaries.

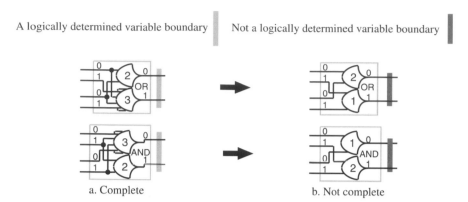

A logically determined variable boundary   Not a logically determined variable boundary

a. Complete   b. Not complete

Figure 2.17   2NCL AND and OR expressions that do not enforce the completeness criterion.

If these expressions are substituted in the half-adder expression on Figure 2.18*a*, the result is the expression of Figure 2.18*b*, which contains simpler logical elements. This circuit preserves the completeness criterion for the circuit as a whole, but the logically determined variable boundaries are farther apart and the orphan paths are longer. So more precise assumptions are required about delay relationships. The reduced logical expressivity necessitates more expression of temporal relationships to completely characterize the behavior of the expression.

2.4.2   No Logically Determined Boundaries in Data Path

One might wish to compromise the logical completeness to the point where there are no logically determined boundaries at all in the combinational expressions of the data path. If there are no logically determined completeness boundaries and the completeness criterion is not in play, then a timing interval must be associated with the expression. One might as well use the less expressive Boolean logic.

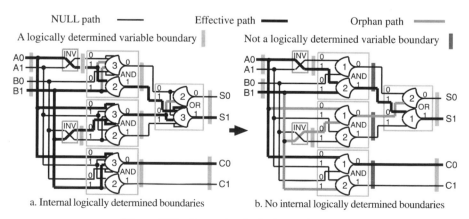

NULL path ——   Effective path ▬▬   Orphan path ▬▬

A logically determined variable boundary ▐   Not a logically determined variable boundary ▐

a. Internal logically determined boundaries   b. No internal logically determined boundaries

Figure 2.18   Lessening logical completeness.

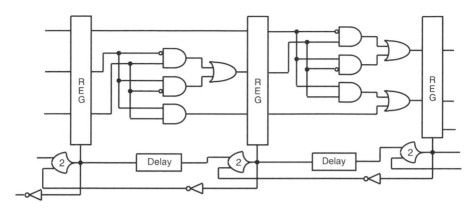

Figure 2.19   Bundled data/micropipeline style expressions.

One might still use logically determined expressions to manage the data path. A delay element can be associated with each combinational expression, and sequencing of these delays can be managed by inserting the delays in a pipeline with a 1 value variable data path, as shown in Figure 2.19, forming a micropipeline or bundled data structure [52,57]. The logically determined variable boundaries of wavefronts flowing through this pipeline and through the delay elements provide the timed propagation boundaries for the combinational expressions coordinating the flow of data through the data path.

2.4.3   No Logically Determined Boundaries at All

The next compromise of logical expressivity would be to remove all logically determined variable boundaries from the system. Every expression in the system posses an associated time interval, and it is convenient if they are all equal and in phase. A single global signal distributed to each combinational expression expressing a single time period will suffice. The result is clocked Boolean logic shown in Figure 2.20. This is the form of least logical determination and most timing assumption.

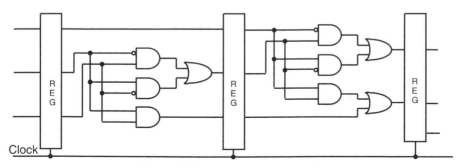

Figure 2.20   Clocked Boolean logic.

2.5 SUMMARY

The search for a sufficiently expressive practical logic was begun by enhancing Boolean logic to sufficient expressivity. This required a 4 value logic (4NCL) to retain the function form of the logical operators. From this was derived a 3 value logic (3NCL) with logic operators that are state holding. From 3NCL one can derive either Boolean logic, which abandons logical completeness, or 2NCL, which preserves logical completeness with a noncritical timing relationship in the orphan path. In 2NCL there is discovered a sufficiently expressive practical logic suitable for electronic implementation.

CHAPTER 3

The Structure of Logically Determined Systems

The structure of a logically determined system is quite different from a clock-driven system. While a clocked system derives its liveness and coordination of information flow from the pulse of the clock, a logically determined system must derive these properties of behavior solely from its logical relationships.

It is first shown that there is a simple logical relationship that is inherently live. It is next shown that this inherently live logical relationship can be evolved into structures called cycles, which can be composed into pipelines that spontaneously propagate data wavefronts. These pipelines can then be composed into structures that coordinate the flow and interaction of data wavefronts implementing complex digital systems.

3.1 THE CYCLE

There is one very simple logical relationship that is spontaneously active.

3.1.1 The Ring Oscillator

A ring oscillator shown in Figure 3.1 is a closed signal path with an odd number of inversions propagating a binary signal. It will continually strive to oscillate and is spontaneously active.

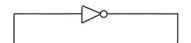

Figure 3.1 Ring oscillator.

Logically Determined Design: Clockless System Design with NULL Convention Logic^{TM}, by Karl M. Fant
ISBN 0-471-68478-3   Copyright © 2005 John Wiley & Sons, Inc.

3.1.2   Oscillator Composition with Shared Completeness Path

Two ring oscillators can be combined and synchronized by sharing a completeness path. The shared completeness path requires that the signal transition in each direction be complete in all paths before the signal transition continues in any path. All shared paths must transition to 1 before the transition to 1 continues through any of the paths. All shared paths must transition to 0 before the transition to 0 continues through any of the paths.

A 2NCL N of N operator expresses a shared completeness path as shown in Figure 3.2. The combined structure will oscillate at the frequency of the slowest oscillator. A fast oscillator will always wait for the transition on a slower oscillator at the completeness path. This structure, called combinational synchronization, synchronizes the signal behavior of the oscillators in phase by combining the signals through a shared completeness path.

When an oscillator shares a different completeness path with each of two other oscillator, as shown in Figure 3.3, it will synchronize through one of its shared completeness paths with one oscillator, and then it will synchronize through the other shared completeness path with the other oscillator. As the individual oscillator spontaneously oscillate, this out-of-phase synchronization results in transition wavefronts propagating from oscillator to oscillator spontaneously flowing through the structure as shown in Figure 3.4. This is the familiar pipeline structure and will

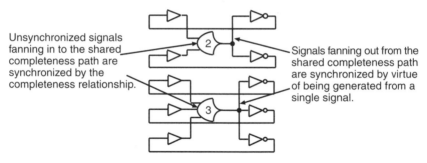

Figure 3.2   Synchronized cycles.

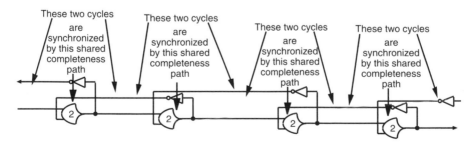

Figure 3.3   Pipeline synchronized cycles.

Time

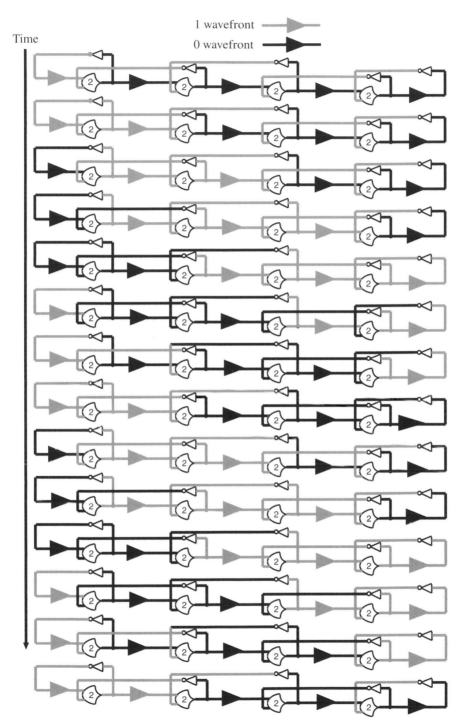

Figure 3.4 Transition wavefronts propagating through pipeline synchronized cycles.

be called pipeline synchronization or simply a pipeline. The behavior of the pipeline is completely logically determined by the shared completeness paths.

The oscillators of the pipeline shown in Figure 3.4 with closed paths on either end will continuously oscillate, and transition wavefronts will continuously flow through the structure. This spontaneous wavefront flow through spontaneously active pipelines composed of coupled oscillators continually striving to oscillate is the foundation of behavior for logically determined systems.

3.1.3   Cycles and 2NCL Data Paths

A 2NCL data path monotonically transitions between complete DATA (one DATA value per variable) and complete NULL (all NULL values). This monotonic transitioning embodies the behavior of a single binary signal.

Figure 3.5 shows a pipeline with a data path of two 4 value variables. The rank of 1 of 4 operators detects the one DATA and all NULL state of each variable and the 2 of 2 operator in the acknowledge path detects the completeness of DATA and NULL over the two variables in the data path. The detection across a data path of completion of DATA and completion of NULL reduces the behavior of a whole data path to a binary signal with the values 'complete data' and 'complete NULL'. Consequently a portion of the binary signal path of a ring oscillator can be replaced with an entire 2NCL data path. The resulting structure, which will be called a cycle, still behaves like a ring oscillator striving to oscillate.

The binary signal from the completion detection forms an acknowledge signal, which is inverted and presented to a previous portion of the data path forming a shared completeness path spanning the entire data path. The spanning completeness

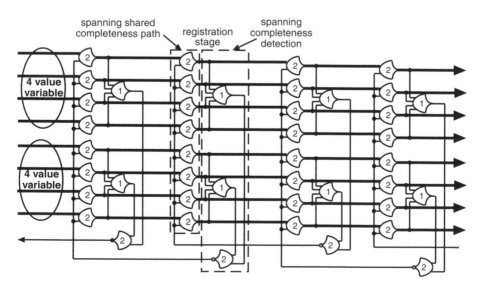

Figure 3.5   Pipelined cycles with 2NCL data path.

detection and the spanning shared completeness path together synchronize the entire data path and condition its behavior to that of a binary signal. The completion detection, the acknowledge signal, its inversion, the spanning shared completeness path, and the portion of the data path between the shared completeness path and the completeness detection form a cycle. A shared completeness path followed by completion detection forms a registration stage synchronizing two cycles.

Transition wavefronts flowing through a pipeline of synchronized cycles are now propagating real data. Arbitrarily wide data paths can be incorporated into cycles and the behavior of the resulting structure remains spontaneously active and logically determined.

3.1.4   Data Path Abstraction

Conversely, since the data path transitions monotonically just like a single binary signal, the entire data path can be abstracted to a single binary signal as shown in Figure 3.6 without changing the synchronization behavior of the cycles. Completion on the data path is trivial, and the shared completion path is a single M of M operator. With this collapsed data path model the behavior of an entire system can be modeled and emulated with single signal data paths to determine the correctness of the system control structure and dynamic behavior issues such as maximum throughput and throughput bottlenecks independent of the data path functionality. These circuits can be simulated, or they can be physically implemented. The structure and behavior of a significant sized system could easily be mapped into a single FPGA, for instance.

Figure 3.6 also illustrates viewing the cycle in terms of a data path part and an acknowledge path part. The acknowledge signal is the portion of a cycle that is actually a binary signal. The figures will follow the convention of illustrating the data paths as wide lines and acknowledge paths as narrow lines.

The discussion of cycle structures will continue in terms of multi-variable data paths, collapsed data paths, and also the graphic shown in Figure 3.7, which indicates a registration stage on an arbitrarily wide data path. The three representations will be used as convenient.

3.1.5   Composition in Terms of Cycles

1. Every signal path in a system must be part of a cycle.

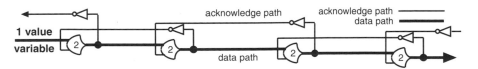

Figure 3.6   Pipeline cycles with data path collapsed into single-value variable data path.

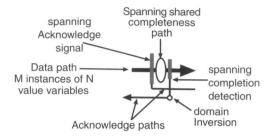

Graphical pipeline of unspecified data path width

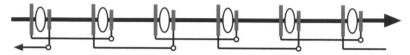

Figure 3.7 Graphical representation of abstract cycle structures.

2. Every cycle must have an odd number of inversions. It is convenient to assume that each cycle has a single inversion in the acknowledge path.
3. Shared cycle paths must be completeness paths.
4. All signals presented to a completeness path must be in the same monotonic transition phase.

3.1.6 Composition in Terms of Registration Stages

1. Each registration stage must acknowledge every registration stage that contributed to its data path.
2. Each registration stage must be acknowledged by every registration stage to which it contributed a data path.

3.2 BASIC PIPELINE STRUCTURES

Cycles compose into pipelines, and pipelines can be composed into complex structures by means of a few basic pipeline structures.

3.2.1 Pipeline Fan-out

One pipeline can fan out to two or more pipelines. The data path is fanned out from the source pipeline to the destination pipelines. Each destination pipeline must acknowledge the source pipeline. The acknowledge signals are collected through a shared completeness path to the source pipeline forming a cycle. When a data path fans out, the associated acknowledge paths fan in. The fan-out cycle, highlighted with shading in Figure 3.8, has one input shared completeness path and two output shared completeness paths.

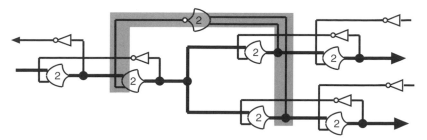

Figure 3.8   Cycles and shared completeness paths of a fan-out pipeline structure.

Cycles will be indicated by highlighting the acknowledge path of the cycle with shading. The cycle also includes that data path between where the spanning completeness generates an acknowledge path and where the acknowledge path spans the data path. Highlighting just the acknowledge path is visually simpler, avoiding the sometimes large structures of the data path.

3.2.2   Pipeline Fan-in

Two or more pipelines can fan-in to a single pipeline through a shared completeness data path. The destination pipeline acknowledge is fanned out to each source pipeline. When data path fans in, the associated acknowledge path fans out. The fan-in cycle shown in Figure 3.9 has two input shared completeness paths and one output shared completeness path. The two input paths are synchronized into the output path by a combinational shared completeness path in the data path.

The fan-in structure is the structure that supports combinational functions. The 2 of 2 operator in the data path of Figure 3.9 can be a 2NCL combinational expression for an arbitrarily wide data path, as shown in Figure 3.10. As long as the input paths merge through a combinational expression enforcing the completeness criterion, they will be properly synchronized into the output path.

3.2.3   The Pipeline Ring

A pipeline ring is a pipeline with its ends connected forming a continuous ring of cycles. There must be at least three cycles in a pipeline ring, as shown in

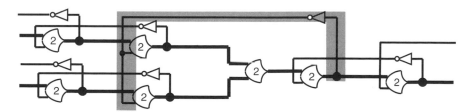

Figure 3.9   Cycles and shared completeness paths of a fan-in pipeline structure.

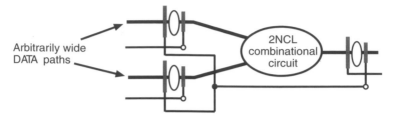

Figure 3.10   Fan-in through a combinational expression shared completeness path.

Figure 3.11. A ring can be combined with other pipeline structures through a shared completeness path as shown in Figure 3.12.

3.2.4   Cycle Structure Example

Cycles and pipeline structures will be summarized with the example of a bit-serial full-adder that features a cycle with three input shared completeness paths fanned in through a combinational expression shared completeness path that fans out to two output shared completeness paths and a three-cycle ring structure feeding the carry-out variable back to the input. The adder, from the point of view of the registration stages, is shown in Figure 3.13a. Figure 3.13b shows the corresponding 2NCL circuit for the serial adder with 2 value variables.

   The full adder is a combinational expression with three 2 value variable sources and two 2 value variable destinations. From the point of view of registration stages each source registration stage must be acknowledged by each destination registration stage it contributes to and each destination registration stage must acknowledge each source registration stage that contributes to it. So both destination registration stages must acknowledge all three source registration stages. The data path has a fan-in of three through the combinational expression and a fan-out of two, so the acknowledge path has a fan-in of two through the 2 of 2 operator and a fan-out of three. The carry feedback ring, shown as a dashed data path, is composed of three cycles.

   Figure 3.14 illustrates the serial adder from the point of view of the cycle structure. The full-adder cycle has three input shared completeness paths and two output shared completeness paths shown in Figure 3.14a. The full-adder cycle is one cycle of the carry feedback ring. Figure 3.14b and c shows the other two cycles.

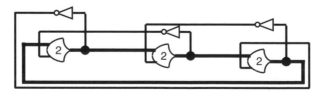

Figure 3.11   A three-cycle ring.

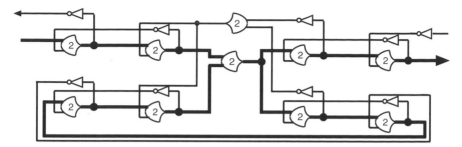

Figure 3.12   Ring synchronized with pipeline.

3.3 CONTROL VARIABLES AND WAVEFRONT STEERING

Spontaneously flowing data path wavefronts are steered through the pipeline structures of a logically determined system by interacting with control wavefronts.

3.3.1 Steering Control Variables

There are no control signals in a logically determined system. A control variable is just another wavefront in the system flowing through a pipeline. It is no different from data wavefronts flowing through data path pipelines. A control variable is typically a multi-value variable, with each value having a unique control meaning. In the figures, a control variable will be shown as gray lines representing the individual values of the control variable. Figure 3.15 shows the graphic and corresponding 2NCL expression for two registration stage control structures with an enabling control value. In Figure 3.15*a* the control value is presented directly to the shared completeness path of the registration stage. Since both the control value and the acknowledge path span the shared completeness path, they can be precombined, as shown in Figure 3.15*b*, into a single value presented to the shared completeness path. In both cases the control value is required for completeness of input and the data path wavefront will be enabled through the registration stage only when the control value is DATA.

Because of the completeness criterion, control wavefronts and the data path wavefronts being controlled are sequence synchronized. The Nth control wavefront arriving at the control structure interacts with the Nth data path wavefront arriving at the control structure. This sequence synchronization is the basis of managing wavefront flow through a system.

3.3.2 Fan-out Wavefront Steering

A 1 to 3 fan-out steering structure is a cycle with 2 input shared completeness paths, the data path and the control path and 3 output shared completeness paths. The values of the 3 value control variable are distributed, one value each, over the

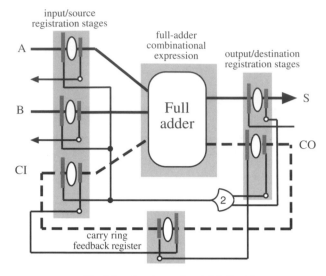

a. Register structure of serial full adder

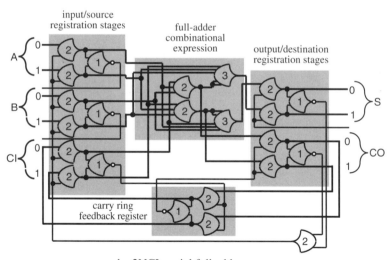

b. 2NCL serial full adder

Figure 3.13   Bit serial full adder.

output paths. Since for each control wavefront only one value of the control variable will be DATA, only one output path will achieve input completeness and pass the data path wavefront from the input path to the enabled output path. Every Nth control path data wavefront will steer every Nth data path data wavefront.

The fan-out cycle has a data path fan-in of two and a data path fan-out of three, so its acknowledge path has a fan-in of three and a fan-out of two. Because only one

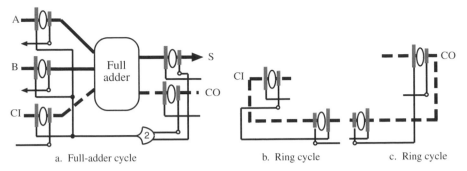

a. Full-adder cycle   b. Ring cycle   c. Ring cycle

Figure 3.14   Cycle structure of serial adder.

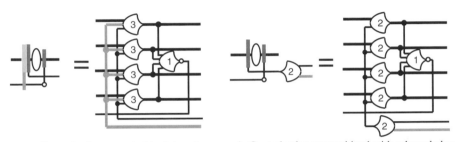

a. Control value presented to data path   b. Control value precombined with acknowledge

Figure 3.15   Registration stages with control values.

output path at a time will ever pass a DATA wavefront, there will be exactly one acknowledge from one output registration stage for each input wavefront. So the acknowledge paths from the output paths are fanned-in through a 1 of 3 operator.

The fan-out steering structure is shown in Figure 3.16 in graphic representation. The control variable is shown in gray with its individual value paths. Figure 3.16a shows each control value spanning the shared completeness path of each output registration stage. Figure 3.16b shows the control values combined with the acknowledge signal. A single signal then spans the shared completeness path.

Fan-out Steering to Multiple Paths. It is possible that a single control value may steer a wavefront to more than one output path. If the path configurations can be uniquely discriminated, then the acknowledge paths can remain mutually exclusive and unambiguous. This is the case with Figure 3.17a. Each control value steers wavefronts to two output paths AB or BC. Unique acknowledge wavefronts can be generated from each steering configuration, so there is no ambiguity of behavior.

The control configuration of Figure 3.17b, however, which steers to A or AB or C cannot directly generated unambiguous acknowledge wavefronts. When A or C is

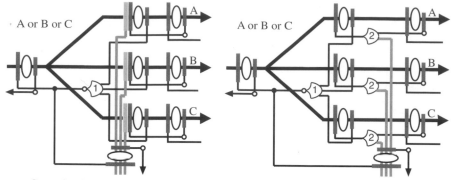

a. Control values spanning the data path   b. Control values combined with acknowledge path

Figure 3.16   Baseline fan-out steering structure, DEMUX.

selected, there is only one acknowledge wavefront, but when AB is selected, there are two uncoordinated acknowledge wavefronts, one from A and one from the combination of A and B. One solution, shown in Figure 3.18a, would be to replicate the registration stage for the A path so that a different stage can be selected for each condition and will generate unique acknowledge wavefronts for each case. A more efficient solution is to condition the acknowledge wavefronts with the control wavefronts shown in Figure 3.18b. Extending the control values into the acknowledge path may at first appear suspicious, but it is perfectly legitimate.

Cycle Analysis of Acknowledge Conditioning. The configuration of Figure 3.19a is the collapsed data path model for the control values conditioning the acknowledge wavefronts of Figure 3.18b. Path **y** of the control value enables the registration stage and path **x** conditions the acknowledge. The first question is whether the two paths are paths of a cycle. The first step is to determine the closed path of the path segment of interest. Beginning on a point of the path segment

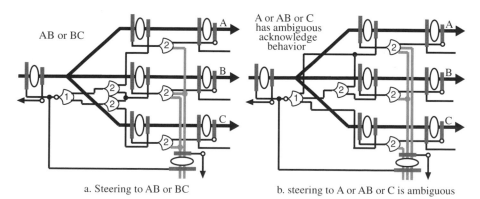

a. Steering to AB or BC   b. steering to A or AB or C is ambiguous

Figure 3.17   Steering a wavefront to multiple paths.

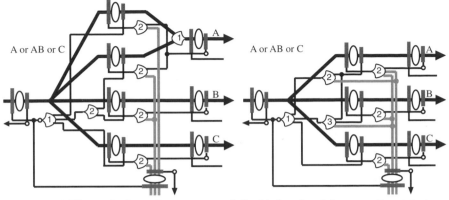

a. Adding a redundant register stage   b. Conditioning acknowledgements with control values

Figure 3.18   Disambiguating acknowledge behavior.

and tracing signal flow, find the path that returns to the beginning point. Figure 3.19*b* shows the closed path of the **x** branch and Figure 3.19*c* shows the closed path for the **y** branch. Both closed paths have a single inversion and are cycles. At the re-convergence of the two branches at the 2 of 2 operator, the two signals remain in the same monotonic transition phase and will obey the completeness criterion so the behavior of the configuration remains logically determined. Notice that if the inversion is before the 2 of 2 operator in the acknowledge path of cycle B, then the **x** branch is in a closed path with no inversion; it is not a part of a cycle, and the re-converging branches are not in the same monotonic transition phase. In this case, putting an inverter in the **x** path would make **x**'s closed path a cycle and would bring the monotonic transitioning of the two branches back into phase.

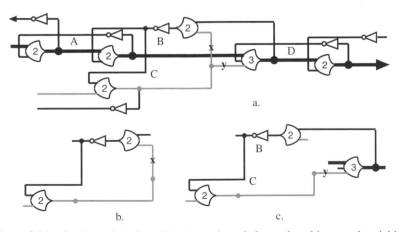

Figure 3.19   Cycle model of conditioning acknowledge paths with control variables.

The Cycles of the Fan-out Steering Structure. Figure 3.20*a*, *b*, and *c* shows the cycle structure enabled by each control value, and Figure 3.20*d* shows the combinational expression of the acknowledge path that conditions the acknowledge signals.

3.3.3 Fan-in Wavefront Steering

A 3 to 1 fan-in steering structure is a cycle with four input shared completeness paths, three data paths and the control path, and one output shared completeness path. The values of the control path variable are distributed, one value each, over the input-shared completeness paths. Since for each control wavefront only one value of the control path variable will be DATA, only one input path per wavefront will achieve input completeness and pass a data wavefront to the output. Every *N*th control path data wavefront will enable an *N*th data path data wavefront through the fan-in cycle.

The fan-in steering structure is shown with spanning control values in Figure 3.21*a* and with precombined control values in Figure 3.21*b*. The control variable is shown in gray with its individual value paths.

Only one input path at a time will be enabled so the data path fans-in to a 1 of 3 operator. But the 1 of N operator is really just the simplest possible combinational

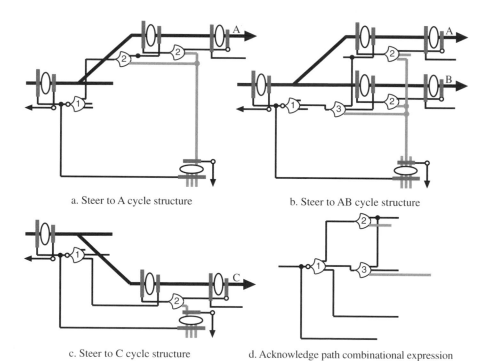

a. Steer to A cycle structure   b. Steer to AB cycle structure

c. Steer to C cycle structure   d. Acknowledge path combinational expression

Figure 3.20   Cycles of the A or AB or C steering structure.

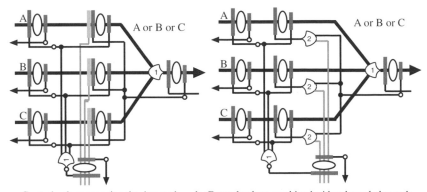

a. Control values spanning the data path   b. Control values combined with acknowledge path

Figure 3.21   Baseline fan-in steering structure, MUX.

expression. If the paths are fanned-in to a combinational expression that can receive multiple input wavefronts, then the fan-in control variable will enable multiple fan-in paths. The most common situation is where 2 paths are always enabled or 3 paths are always enabled, as in Figure 3.22a, which enables AB or BC, and completeness relationship can be unambiguously expressed. But there can also be a varying number of paths enabled as in Figure 3.22b, which enables A or AB or C, and in this case completion behavior is ambiguous for both the combinational expression and for the acknowledges to the control variable path. The acknowledges can be conditioned with the control values, as in Figure 3.22b, just as they were with the fan-out steering structure of Figure 3.18b, but the combinational expression must also be conditioned.

When a completeness path always receives a constant number of variables, the determination of completeness is the number of DATA values equal to the

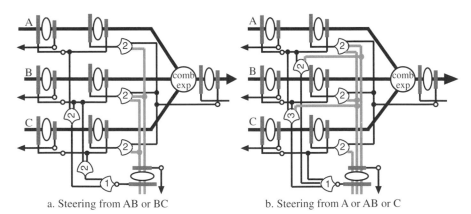

a. Steering from AB or BC   b. Steering from A or AB or C

Figure 3.22   Multi-path fan-in to a combinational expression.

number of variables. But when a path can receive a different configuration of variables for each different instance, the determination of completeness must include a specification of which configuration of variables is involved. Typically there is a control variable that determines the configuration of variables, and that determining variable can be used to condition the completeness relationships.

In the example the combinational expression can receive A or A and B or C and cannot, on its own terms, determine which configuration is presented. The fan-in control variable determines the configuration, and the fan-in control variable must be used to condition the completeness behavior of the combinational expression.

The first inclination might be to simply extend the control variable into the combinational expression as in Figure 3.23 in the same manner that the control variable was extended into the acknowledge path. This, however, does not work.

To understand what is incorrect and how to make it correct, one must consider the cycle structures. Remember that every path in a system must be a path of a cycle, that a cycle must have an odd number of inversions, and that signals presented to a completeness path must be in the same monotonic transition phase. Figure 3.24 shows the cycle model for the fan-in structure corresponding to Figure 3.23. The control value path has three branches. Branch **y** enables the source paths. Branch **x** conditions the acknowledge paths, and branch **z** conditions the combinational expression. If for each branch path one traces the closed path, the closed path will be a cycle or it will not be a cycle. The closed paths for each branch of the control variable are shown in Figure 3.24b, c, and d. The closed paths for branch paths **x** and **y** form a cycle with a single inversion. The closed path for branch path **z** has two inversions. Branch path **z** is not a path of a cycle.

The solution shown in Figure 3.25a is to pipeline the control variable one more cycle before presenting it to the combinational expression of cycle D. Path **z** is divided into two paths **z1** and **z2** by inserting a shared completeness path creating

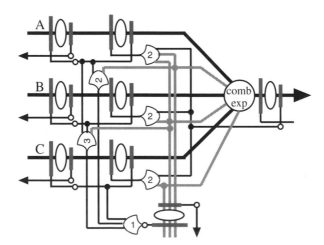

Figure 3.23 Incorrect presentation of the control variable to the combinational expression.

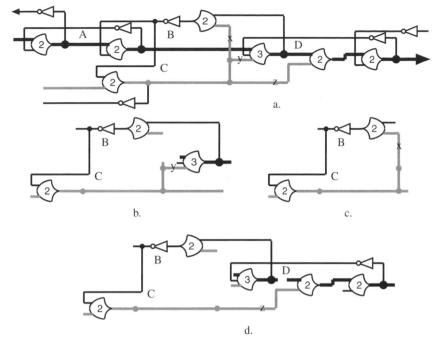

Figure 3.24   The cyclic paths of the control variable.

a new cycle. The individual closed paths for **z1** and **z2**, are now cycles, as shown in Figure 3.25*b* and *c*. The correct fan-in structure is shown in Figure 3.26.

3.3.4   Wavefront Steering Philosophy

Logically determined systems are fan-out control oriented in contrast to the fan-in control orientation of clocked Boolean systems. Figure 3.27 shows a fan-out to

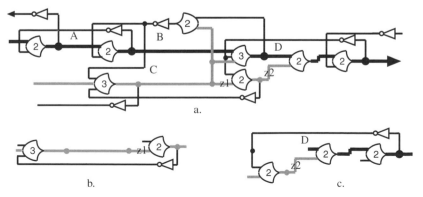

Figure 3.25   The control variable extended with a pipeline stage.

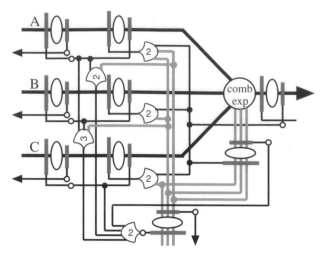

Figure 3.26   A or AB or C fan-in structure with control variable pipelined to combinational expression.

function units then a result fan-in structure. Wavefronts are steered from the source registration stage to the selected function unit. The wavefront propagates only through the data path of the selected function unit. The other two data paths remain NULL and quiescent. Exiting the function unit, the wavefront flows through the fan-in node with no interference from the other two paths, which remain NULL, and the destination stage acknowledges the steering stages, which allow NULL to pass resetting the selected function unit path. The destination stage receives the NULL wavefront and acknowledges the data path and control path to allow the next data wavefront, which will be steered by the next control wavefront.

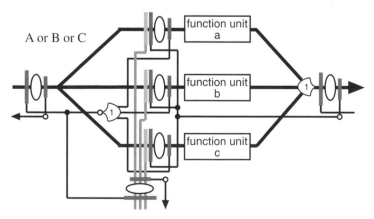

Figure 3.27   Fan-out (DEMUX) oriented function select control.

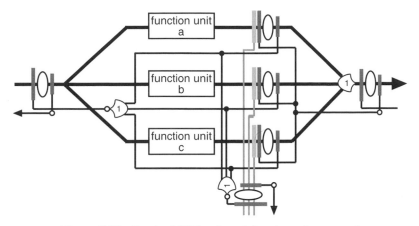

Figure 3.28 Fan-in (MUX) oriented function select control.

Figure 3.28 shows a postfunction fan-in oriented control structure. In this case the wavefront from the source registration stage flows through all three function unit data paths, and all three assert a wavefront to the fan-in structure. The fan-in control selects one of the wavefronts to be allowed through to the destination registration stage. The fan-in control acknowledges the source registration stage, which allows a NULL wavefront to pass and flow through all three function unit data paths.

There are two reasons why not to use this structure for logically determined system design. The first is that the paths through the unselected functions are ineffective data paths that are very long orphan paths. This violates the strategy of minimizing and isolating orphan paths and compromises the logical determinability of the system. In the fan-out select structure the orphan paths are isolated to the wire branches from the source stage to the selection registration stages. The second is that flowing wavefronts through the unselected functions is an unnecessary waste. Two of the propagated wavefronts get thrown away.

In a logically determined system wavefronts are steered through effective data paths, and there should only very rarely be circumstances where a propagated wavefront is thrown away. One such circumstance might be with speculative resolution where two results are purposely precomputed and a later one is selected while the other is thrown away, such as in a carry select adder.

So in logically determined design it makes sense to pre-steer wavefronts at fan-out and makes no sense to postselect wavefronts at fan-in. Clocked Boolean design is just the opposite. Boolean logic functions will always output data regardless of whether or not they received valid data. At a result fan-in node one data set must be selected. Preselecting at fan-out is logically meaningless. One might preselect to avoid data path switching and save power but all three functions will still assert data into the fan-in node and one must be selected.

One might say that while clocked Boolean design is inherently MUX oriented, logically determined design is inherently DEMUX oriented.

3.3.5 Concurrent Pipelined Function Paths

While one should always steer with fan-out select, there are circumstances where fan-in select is necessary. For instance, if the function paths are pipelined, the exclusivity of wavefronts arriving at the fan-in node is not preserved. Wavefronts can be launched sequentially into multiple function paths and may arrive at the fan-in node out of sequence. They must be resequenced into the fan-in node by a fan-in control structure as shown in Figure 3.29. This can be accomplished by pipelining the fan-out control to the fan-in control. The Nth fan-out control wavefront will steer the Nth data wavefront into the function structure and the same Nth control wavefront will steer the Nth data wavefront out of the function structure through the fan-in node. The data wavefronts are correctly sequenced through the structure by the control wavefronts. The structure supports opportunistic concurrent activity among the functions units as well as the pipeline concurrency within the function units.

3.4 THE LOGICALLY DETERMINED SYSTEM

The purpose here is to give the reader an intuitive sense of the structure and behavior of a logically determined system, which is quite different from the structure and behavior of the familiar clocked synchronous system. At its most essential a logically determined system is a complex structure of coupled oscillators. At a higher level of abstraction it is a structure of pipelines through which Information flows via spontaneously propagating wavefronts that are steered, that interact, that combine and split, and that transform. Control wavefronts and data path wavefronts flow quite autonomously through the system. There is no subdomain of control. There are just wavefronts flowing through the system, and that's all there is, period.

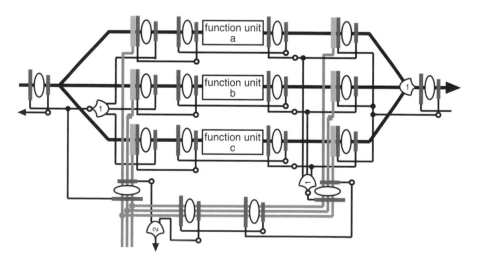

Figure 3.29 Structure with independent paths requiring fan-in control.

There are no signals in the familiar sense. All signal paths in the system are enlisted to implement cycles. Cycles are enlisted to implement pipelines, pipelines are structured into a system, and wavefronts flow through the pipelines.

3.4.1 Managing Wavefront Interaction

Managing wavefront interaction is a matter of how the wavefronts enter the pipelines of the system and how they interact. An essential aspect of the system structure is the completeness criterion of each shared path in the system. Wavefronts interact at shared paths. When an Nth wavefront arrives at a shared path, it will be blocked and will wait until an Nth wavefront arrives on the other path. When all Nth wavefronts are present at a shared path, they will form completeness and will interact. The result wavefront will flow to the next shared path and wait for another wavefront or interact with the other wavefront already waiting. This waiting for completeness, which is fundamental to the logical design methodology right down to the logic operators, is the coordination mechanism of the system.

There are no control signals. There is no global or local time reference. While the flow of wavefronts through each cycle is strictly sequential, the flow of wavefronts through the system as a whole is generally concurrent. In a behaving system there is no state anywhere in the system that is reliably samplable at any given instant. The state of the system resides in the flowing wavefronts, and there is no way to determine when wavefronts might be interacting and transforming or when they might be stable. From the point of view of the system state the behaviour of the system appears to be chaotic and indeterminate. The behavior of a logically determined system must be viewed in terms of its logically relationships, and in this context its behaviour appears quite orderly and completely determined.

3.4.2 A Simple Example System

Consider a simple system modeled, for the purpose of familiarity to the reader, on a traditional sequential processor architecture shown in Figure 3.30. It has a memory (M), a function unit (F), a register file (R), and an instruction decoder (I). The memory, the function unit, and the register file are all connected by black data paths with steering elements. The instruction decoder is connected to all the other units with gray control variable pipelines. Again, they are not control signals. They are pipelines through which control variable wavefronts flow. There is no difference between a data path and a control variable path except labeling. They are both pipelines built of cycles.

Each instruction in sequence is decoded by the instruction decoder into several control variables, each of which flows into its own control pipeline. The Madd/ op pipeline has two variables, the memory address variable and a 2 value variable with the values R and W for read and write. The MFOsel pipeline controls the fan-out structure for the memory read operation and is a single variable with the values MI and MR. The RFOsel pipeline controls the fan-out structure for the register file read operation and is a single variable with the values RM, RFA, RFB, and

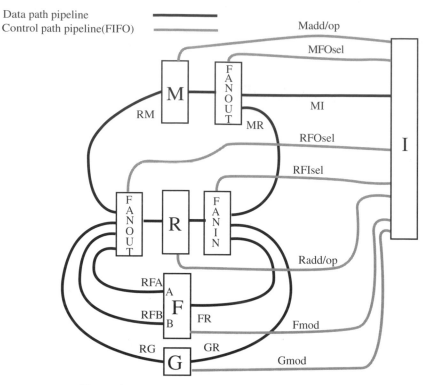

Figure 3.30 Pipeline structure of simple processor.

RG. The RFIsel pipeline controls the fan-in structure for the register file write operation and is a single variable with the values MR, FR, and GR. The Radd/op pipeline has two variables, the register address variable and a 2 value variable with the values R and W for read and write. The Fmod pipeline is the function modifier for the F function unit and is a single variable with the values f1, f2, and f3. The Gmod pipeline is the function modifier for the G function unit and is a single variable with the values g1 and g2.

The behavior of the system begins by initializing it to stable state, and initiating an instruction fetch by placing an address wavefront in the Madd/op pipeline and a select wavefront in the MFOsel pipeline. The address wavefront flows through the memory (M) and is transformed into a data wavefront that flows to the FANOUT and, interacting with the MFOsel wavefront, is steered to the instruction decoder (I). Flowing through the instruction decoder, the data wavefront is split into field wavefronts conditioned by the OP code field wavefront. Each field wavefront, as it flows, is encoded or decoded to form its appropriate control wavefront and thus flows to its appropriate control pipeline. This first fetched instruction reads a location from memory and places it in the register file (R). The instruction decodes to a memory address (MAD/op), a register address (Radd/op), a memory FANOUT select (MFOsel), and a register FANIN select. (RFLsel) All four of these wavefronts are sent into their pipelines together. The memory address wavefront transforms into a data wavefront that interacts with the MFOsel wavefront, flows to the register file FANIN, and interacts with the RFLsel, which enables it to flow to the register file where it interacts with Radd/op wavefront and is stored.

Say all traffic to the register file is blocked and that the pipelines are backing up. Consider that 5 consecutive instructions read from the memory and write to the register file. There will be 10 read operations to the memory. Five reads will be instructions fetches steered to the instruction decoder and five will be data reads that are steered to the register file and that will back up in the pipeline. There will also be 5 RFIsel wavefront issued by the instruction decoder that will back up in the RFIsel pipeline. When the register file starts accepting writes, the data wavefronts and the control wavefronts will flow from their pipelines in sequence. The first data wavefront will be steered by the first RFIsel wavefront, the second by the second, and so on. As long as the instruction decoder is placing the control wavefronts into their pipelines in the sequence that the instructions generated them, all wavefronts flowing through the system will correctly meet at shared completeness paths and properly interact.

Figure 3.31 shows a more extensive slice of system activity for a short selection of simple instructions. There are load from memory to register file instructions (LD), function F instructions (F), function G instructions (G), and store to memory from the register file instructions (ST). For each decode step the decoder initiates control variable wavefronts into the control path pipelines. All the control variables in the control path pipelines are in instruction order. Figure 3.31 shows the decode steps and the value of the control variable wavefronts initiated at each step. For the function instructions there are multiple decode steps. One instruction fetch is shown at the beginning just to show how it works. The decoder initiates the control wavefronts, and they begin propagating autonomously through the control pipelines and through the system.

instruction	Madd/op	MFOsel	RFOsel	RFIsel	Radd/op	Fmod	Gmod
fetch inx	inx/R	MI					
LD 1,add1	add1/R	MR		MR	1/W		
LD 2,add2	add2/R	MR		MR	2/W		
LD 3,add3	add3/R	MR		MR	3/W		
F 1,2,4,f2			RFA		1/R	f2	
			RFB		2/R		
				FR	4/W		
F 2,3,5,f3			RFA		2/R	f3	
			RFB		3/R		
				FR	5/W		
G 5,6,g1			RG		5/R		g1
				GR	6/W		
ST 5,add4	add4/W		RM		5/R		
ST 6,add5	add5/W		RM		6/R		

control variables generated by decoder into each control pipeline/FIFO

Figure 3.31   Decoder activity.

It will be assumed here, for purposes of illustration, that the instruction decode is very fast and the pipelines are very slow, with the result that the control wavefronts back up in the pipelines. Pipelines naturally act as FIFOs. The instructions are decoded in sequence. The decoded control variable wavefronts enter the FIFOs in sequence, and each FIFO maintains the sequence of wavefronts presented to it.

Figure 3.32 shows the contents of the pipeline/FIFOs after the decoder has decoded all the instructions. The top of the diagram is the exit of each FIFO. So each value at the top of each FIFO is a wavefront awaiting another wavefront at a shared completeness path, except for the memory and register file read operations that initiate data wavefronts. The g1 cannot flow until a data wavefront arrives at the G function, which will not happen until the 5/R of Radd/op pairs with the RG of RFOsel. Similarly the f2 has to await a data wavefront at the F function. The 1/W in Radd/op is awaiting a data wavefront from the fan-in element. The MR of RFIsel is awaiting a data wavefront from the memory. The RFA of RFOsel is awaiting a data wavefront from the register file, which will not occur until the 1R of Radd/op is effected and so will not occur until the three writes are completed. The MI of MFOsel is awaiting a data wavefront from the memory.

The inx/R of Madd/op is not waiting and will generate a data wavefront from memory. This first data wavefront will be steered by MI to the instruction decoder, but the next data wavefront from add1/R will be steered by MR to the register file and begin the flow of wavefronts that will carry out the resolution of the instruction sequence. As long as the decoder keeps decoding instructions and sending control variable wavefronts into the control path pipelines, the system will continue flowing and correctly resolving instruction streams.

To keep the example simple, the condition code data path from the function units to the instruction decoder has been ignored. A logically determined system works best with deep finely grained pipelines. If the decoder has to frequently wait on the condition code, these pipelines will frequently be idle. While a logically

inx/R	MI	RFA	MR	1/W	f2	g1
add1/R	MR	RFB	MR	2/W	f3	
add2/R	MR	RFA	MR	3/W		
add3/R	MR	RFB	FR	1/R		
add5/W		RG	FR	2/R		
add6/W		RM	GR	4/W		
		RM		2/R		
				3/R		
				5/W		
				5/R		
				6/W		
				5/R		
				6/R		
Madd/op	MFOsel	RFOsel	RFIsel	Radd/op	Fmod	Gmod

contents of each control pipeline/FIFO

Figure 3.32 The contents of each control pipeline/FIFO.

determined system with this added path will operate correctly, it will not operate efficiently. While this has usefully served as a simple and familiar example of a logically determined system, the approaches to system architecture that have been so effective for sequential state-based architectures are not so effective for logically determined system design.

3.5 INITIALIZATION

A logically determined system must begin activity in a unambiguously defined logical state. This typically does not occur with power up of a system, so there must be an explicit initialization operation. Every operator and every signal in the system must be set to a stable logic state prior to beginning activity. The initialization operation is not part of the logic of the system, and it is not logically determined. An initialization signal must force the output of a sufficient number of operators to establish the initial state of the system, and this signal must be held long enough for the initial system state to stabilize. Once this is accomplished the initialization signal can be released, and the system can begin behaving according to its logical relationships.

3.5.1 Initializing the System

One obvious strategy is to initialize every operator in the system. A more efficient approach is to initialize the shared paths of the registration stages to NULL. Since all combinational expressions respond to the universal NULL function, the NULL wavefronts will propagate through the combinational expressions, initializing them to NULL. The combinational expression then need not be explicitly initialized. As shown in Figure 3.33, the data path is set to NULL, the combinational expressions propagate to NULL, and the acknowledge paths all request DATA. When the init signal is released, the expression is ready to receive data wavefronts.

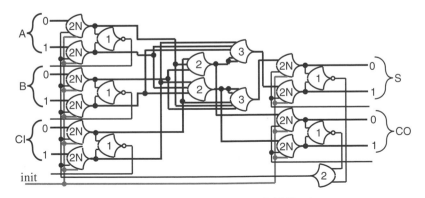

Figure 3.33 Registration stage initialization.

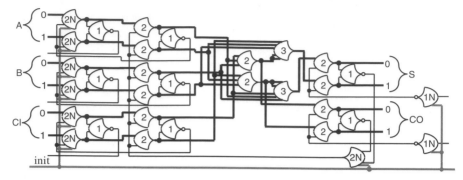

Figure 3.34 Acknowledge path initialization.

A more efficient initialization approach is to initialize the acknowledge paths shown in Figure 3.34. If the data path is initialized to NULL at the beginning of a pipeline and all acknowledge signals are set to request NULL, then the initialized NULL wavefront will propagate through the entire pipeline. The init signal must be held long enough for the NULL wavefront to completely propagate. When the init signal is released, the acknowledge paths will immediately transition to requesting DATA, and the system will be ready to receive DATA wavefronts. Operators that can be initialized are more complex than operators that cannot be initialized. In general, initializing the acknowledge path is the most efficient approach to initialization. If very fast initialization or reset is required, the data path approach is faster.

3.5.2 Initializing Data Wavefronts

Frequently a DATA wavefront must be initialized internal to the system. This initialization can be accomplished with registration stage initialization as shown in Figure 3.35. For the states of the associated acknowledge signals to be properly initialized, the stages surrounding the stage initialized to DATA must be explicitly initialized to NULL.

Initializing a DATA wavefront with acknowledge path initialization, shown in Figure 3.36, is a little more involved but more efficient. The top path of the middle stage is initialized to DATA. Because of the initialized DATA value, the acknowledge signal from this stage will become NULL, so it does not need to be

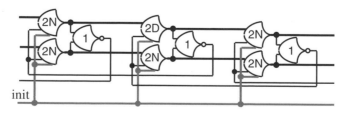

Figure 3.35 Initializing a DATA wavefront.

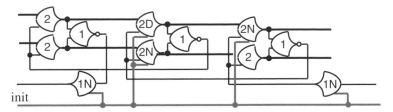

Figure 3.36   DATA wavefront initialization with acknowledge path initialization.

initialized to NULL. The other acknowledge signals will attempt to request DATA and must be initialized to NULL to allow the NULL wavefronts to propagate through the data path. The stage just ahead of the initialized stage, which is presented with a DATA value, must be initialized to NULL. The acknowledge signal from this stage that will acknowledge the stage being initialized is not initialized to NULL but is allowed to request DATA. This is necessary to maintain the initialized DATA value when the init signal is reset. If this acknowledge signal were initialized to NULL, the NULL wavefront presented to the stage would immediately overwrite the initialized DATA value. In the following stage only the paths that receive a DATA value must be explicitly initialized to NULL.

3.6   TESTING

While it appears the behavior of logically determined system is very dynamic and unobservable, there are many characteristics that make testing easy. The first is that the behavior is logically determined. There is no ambiguity about timing relationships causing glitches or intermittent failures. Because of the redundant encoding of the data, a logically determined system will either halt by failing to achieve a completeness relationship somewhere in the system or will exhibit an illegal code. If a system halts, it is not a difficult problem to trace the source of the halt. Illegal data codes (more than 1 DATA value per variable) can be easily checked during operation.

The logically determined behavior of an implemented system is identical to the logically determined behavior of its software simulation. This provides a powerful tool for testing hypotheses about behavior, for verifying a diagnosis, and for analyzing performance.

While a logically determined system in operation is very dynamic, it can be easily placed in static state conditions for testing. Upon enabling or disabling selected output acknowledge signals and then entering selected wavefronts into the system, the wavefronts will back up in the system creating stable static states that can then be queried.

As a system is receiving its initial flow of wavefronts before it settles to a stable flow behavior, internal cycles with different periods influence the output port and the input ports, creating a startup signature of wavefront periods. The same thing occurs when the system is emptying of wavefronts. These wavefronts period signatures can be used in testing and in gaining limited information about the behavior of cycles internal to the system.

There are many advantages in terms of lack of ambiguity of behavior, controllability of behavior, and reproducibility of behavior that make logically determined systems easy to test and debug. Current testing systems designed to operate with clocked systems can even be conveniently used [51].

3.7 SUMMARY

A complex digital system is composed as a structure of pipelines. Pipelines are composed of coupled cycles. Cycles enliven the system by autonomously striving to oscillate, producing wavefronts that spontaneously flow through the pipelines. All of this is expressed entirely in terms of 2NCL. No other form of expression is needed. A digital system is completely expressed and its behavior completely determined solely in terms of logical relationships.

3.8 EXERCISES

3.1. In general, a cycle can have an odd number of inversions anywhere in the path of the cycle. Discuss why it is best to ignore this generality and to limit the notion of a cycle to a single inversion in the acknowledge path.

3.2. Are there other primitive logical expressions that, like the ring oscillator, are spontaneously active and that cannot be characterized as a composition of ring oscillators?

3.3. Define a language of cycle composition.

3.4. Define the procedure to analyze orphan paths and completeness conformance for expressions in NCL.

3.5. Define a procedure to analyze cycle path integrity for an expression in NCL.

3.6. Define the procedure to automatically extract the collapsed data path model from a complete data path expression in NCL.

3.7. Define a procedure for generating a cycle composition expression from a behavioral expression.

CHAPTER 4

2NCL Combinational Expression

A logically determined 2NCL combinational expression, in addition to implementing the combinational function, implements a shared completeness path that synchronizes two or more cycles. There are three issues associated with creating logically determined combinational expressions in 2NCL. The first is the specification of the function itself. The second is the fulfillment of the completeness criterion. If the completeness criterion is not fulfilled, the expression is not logically determined and does not implement a shared completeness path. The third is with isolating the orphan paths. If orphan paths are not isolated, it cannot be said that an expression is, for all practical purposes, logically determined. Long orphan paths can create critical timing issues and require explicit timing analysis to verify the correct behavior of the expression.

In this chapter we do not aim to present a detailed design methodology but rather to give the reader a sense of the territory. In this chapter we cover the derivation of the 2NCL library of operators, basics of 2NCL combinational design, approaches to synthesize 2NCL combinational expressions, and the mapping of 2NCL combinational expressions directly from Boolean logic combinational expressions.

The first topic is the classification of threshold functions and of Boolean functions. This provides a conceptual foundation for the derivation of the 2NCL operator library and for the synthesis of 2NCL combinational expressions.

4.1 FUNCTION CLASSIFICATION

Functions of a particular range, all three input Boolean functions, for instance, may be classified by commonalities of logical structure. The structural elements can be partitioned into elements that are cheap to implement and elements that are costly to implement. For instance, if the internal logic of a function is costly, rearranging the input wires is free and inverting the input signals can be relatively cheap. It would be useful to be able to express a few functions in terms of the internal logic structure and to express the rest of the functions in terms of rearranging and inverting the inputs.

Logically Determined Design: Clockless System Design with NULL Convention Logic^{TM}, by Karl M. Fant
ISBN 0-471-68478-3   Copyright © 2005 John Wiley & Sons, Inc.

Functions can be sorted into classes that differ only in the permutation and/or negation of their input signals, called PN classification. The implementation of a single function from each class as a representative function is then sufficient to implement the entire class of functions by appropriately connecting and inverting input signals of the class representative function. The result is a small group of functions that are generally expressive over the range of all possible functions.

Function classification was developed initially to manage the mapping of Boolean functions to a minimal set of vacuum tube circuits [19]. It was extensively developed in the study of threshold logic to facilitate the synthesis of threshold expressions [36,22]. In the case of 2NCL it provides a rationale of design and allows confident statements of thoroughness that would, otherwise, be difficult to justify.

4.1.1 Threshold Function Classification

The function classes for three input positive threshold functions are derived as an illustrative example of function classification. For this discussion of function classification the state-holding behavior of 2NCL operators will be ignored, and only their threshold function will be considered. Preliminarily Figure 4.1 shows the three classes of 1 and 2 input threshold functions.

The classification process is illustrated in Figure 4.2. All forms of three input functions with input weights of 1 and 2 and all meaningful thresholds are considered. The process begins with a 3 input function with all input weights 1 and threshold 1 at upper left. This enumeration proceeds through all possible functions to all input weights 2 and threshold 6 at the lower right. In the process, functions occur that can be mapped to a previously enumerated function. Each function encountered that cannot be mapped to a previously enumerated function is taken as the representative function of a class and is highlighted with shading. As the enumeration proceeds, each function is either identified as a new class representative function or its class membership is shown by a mapping to its functionally equivalent class representative function. Some of the 3 input functions map to 2 input function classes.

Input weights of 3 are not considered because most input weights of 3 for a three input function can always be reduced to a weight of 2 without changing the functionality. The purpose of the weights is to differentiate the inputs. A weight of 3 in relation to a weight of 1 does not differentiate any more than a weight of 2 in relation to a weight of 1. So, for all combinations of weight 3 and weight 1, the weights of 3 can be reduced to weight 2 with an appropriate adjustment in threshold. For all combinations of weight 3 and weight 2, both weights can be reduced by one with the appropriate adjustment in threshold. The only combination that is not obviously

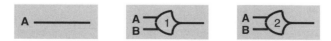

Figure 4.1   Classes of 1 and 2 input threshold functions.

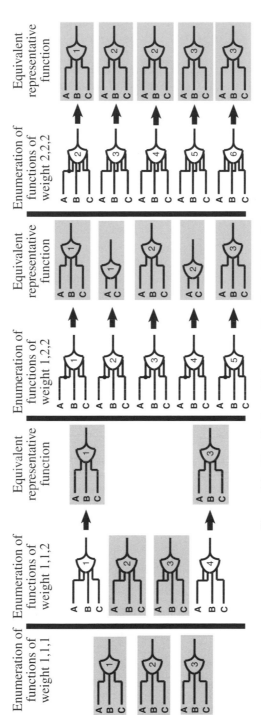

Figure 4.2 Classification of 3 input positive threshold functions.

reducible is a combination of weights 3, 2, and 1. The mapping to equivalent classes for 3 input functions with weights 3, 2, 1 for threshold values through 6 is shown in Figure 4.3.

The procedure discovered five classes of threshold functions of exactly 3 inputs. The five 3 input classes, the two 2 input classes, and the one 1 input class are sufficient to express all possible threshold functions of 3 inputs or less. If the exercise is repeated for 4 input threshold functions, 17 classes will be found. In total, there are 25 threshold function classes for threshold functions of 4 or fewer inputs.

4.1.2 Boolean Function Classification

There are two classifications of Boolean functions that are of interest. The PN (or NP) classification considers all functions that differ only in the permutation and/or negation of input signals as being in the same class. And the NPN classification, which adds negation of the output such that all functions that differ only in the permutation and negation of the inputs and negation of the output are in the same class. NPN classification combines some NP classes resulting in fewer classes overall. Figure 4.4 is a table after Muroga [36] of Boolean function classes of 3 and fewer variables that shows the 22 PN classes and the 14 NPN classes. Each function class is defined by a representative equation. The PN functions in bold type are the classes of unate functions of 3 and fewer inputs.

The statistics for PN and NPN classification of Boolean functions is given in Table 4.1 [22]. It can be seen that the classification issue rapidly grows for functions of 4 and 5 variables. As will be clear shortly, the interest here is in the classes of unate Boolean functions and those classes remain quite tractable through 4 variables.

4.1.3 Linear Separability and Unateness

There is a very direct relationship between Boolean unate functions and threshold functions through the property of linear separability. All threshold functions map to linearly separable Boolean functions. Linear separability is illustrated in Figure 4.5 in two dimensions for 2 variable Boolean functions. If a line can be drawn separating the different result values in the function space mapping as with AND and OR, then the function is linearly separable and can be determined by a threshold operator. If the different result values cannot be separated by a line as with XOR, then the function is not linearly separable and cannot be determined by a threshold operator. For three input functions the linear separation is a plane through a function cube, and for four input functions it is a hyperplane through a function hypercube.

A unate Boolean function is one in which no input variable appears in the function in both complemented and uncomplemented form. The boldface PN function classes in Figure 4.4 are unate, the lightface PN functions are not.

All linearly separable Boolean functions map to threshold functions. All linearly separable Boolean functions are unate, but not all unate Boolean functions are

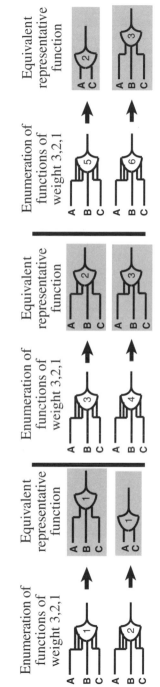

Figure 4.3 Equivalence classes for 3 input positive threshold functions with input weights of 3, 2, and 1.

		PN classes permutation and negation of inputs	NPN classes permutation and negation of inputs and negation of function
0 variable	M0	0	
	M1	1	1
1 variable	M2	x_1	x_1
2 variable	M3	x_1x_2	
	M4	$x_1 \lor x_2$	$x_1 \lor x_2$
	M5	$x_1 \oplus x_2 = x_1\bar{x}_2 \lor \bar{x}_1x_2$	$x_1 \oplus x_2 = x_1\bar{x}_2 \lor \bar{x}_2x_3$
3 variable	M6	$x_1x_2 \lor x_2x_3 \lor x_1x_3$	$x_1x_2 \lor x_2x_3 \lor x_1x_3$
	M7	$x_1 \oplus x_2 \oplus x_3 = x_1(x_2\bar{x}_3 \lor \bar{x}_2x_3) \lor \bar{x}_1(\bar{x}_2\bar{x}_3 \lor x_2x_3)$	$x_1 \oplus x_2 \oplus x_3 = x_1(x_2\bar{x}_3 \lor \bar{x}_2x_3) \lor \bar{x}_1(\bar{x}_2\bar{x}_3 \lor x_2x_3)$
	M8	$x_1x_2x_3$	
	M9	$x_1 \lor x_2 \lor x_3$	$x_1 \lor x_2 \lor x_3$
	M10	$x_1(x_2 \lor x_3)$	$x_1(x_2 \lor x_3)$
	M11	$x_1 \lor x_2x_3$	
	M12	$x_1x_2x_3 \lor \bar{x}_1\bar{x}_2\bar{x}_3$	$x_1x_2x_3 \lor \bar{x}_1\bar{x}_2\bar{x}_3$
	M13	$(x_1 \lor x_2 \lor x_3)(\bar{x}_1 \lor \bar{x}_2 \lor \bar{x}_3)$	
	M14	$\bar{x}_1x_2x_3 \lor x_1\bar{x}_2 \lor x_1\bar{x}_3$	$\bar{x}_1x_2x_3 \lor x_1\bar{x}_2 \lor x_1\bar{x}_3$
	M15	$x_1(x_2x_3 \lor \bar{x}_2\bar{x}_3)$	$x_1(x_2x_3 \lor \bar{x}_2\bar{x}_3)$
	M16	$x_1 \lor x_2\bar{x}_3 \lor \bar{x}_2x_3$	
	M17	$x_1x_2 \lor x_2x_3 \lor \bar{x}_1x_3$	$x_1x_2 \lor x_2x_3 \lor \bar{x}_1x_3$
	M18	$\bar{x}_1x_2x_3 \lor x_1\bar{x}_2x_3 \lor x_1x_2\bar{x}_3$	$\bar{x}_1x_2x_3 \lor x_1\bar{x}_2x_3 \lor x_1x_2\bar{x}_3$
	M19	$x_1x_2 \lor x_2x_3 \lor x_1x_3 \lor \bar{x}_1\bar{x}_2\bar{x}_3$	
	M20	$x_1\bar{x}_2\bar{x}_3 \lor x_2x_3$	$x_1\bar{x}_2\bar{x}_3 \lor x_2x_3$
	M21	$(x_1 \lor \bar{x}_2 \lor \bar{x}_3)(x_2 \lor x_3)$	

Figure 4.4 PN and NPN classes for 3 or fewer variables with representative function equations.

TABLE 4.1   Boolean Function Classification Statistics

Number of Input Variables, N	0	1	2	3	4	5
Total number of all Boolean functions	2	4	16	256	65,536	-4.3×10^9
Total number of Boolean functions of exactly N variables	2	2	10	218	64,594	-4.3×10^9
NP classes of all Boolean functions	2	3	6	22	402	1,228,158
NP classes of Boolean functions of exactly N variables	2	1	3	16	380	1,227,756
NPN classes of all Boolean functions	1	2	4	14	222	616,128
NPN classes of Boolean functions of exactly N variables	1	1	2	10	208	615,904
Classes of all threshold functions		1	2	5	17	
NP classes of unate Boolean functions		1	2	5	20	

AND

OR

XOR

Figure 4.5 Linear separability.

linearly separable [36]. There are 28 NP classes of unate Boolean functions of 4 or fewer variables. Twenty-five of these classes are linearly separable and map directly onto the 25 classes of threshold functions with 4 or fewer inputs.

4.2 THE LIBRARY OF 2NCL OPERATORS

In the interest of completeness and convenience of mapping with Boolean equations, the three 4 variable unate Boolean function classes that are not linearly separable are included in the library as multiple threshold operator expressions. By relying on both threshold function classification and Boolean function classification, one can be confident that the set of operators in the library is a covering set of four input threshold operators that also map to all four variable unate Boolean expressions.

The library of 2NCL operators are shown in Figure 4.6 with a canonical ordering of increasing complexity of threshold functionality. Operators 1 through 25 represent the 25 threshold function classes of 4 or fewer inputs. The 3 multi-operator expressions 26, 27, and 28 represent the three 3 Boolean unate function classes that are not linearly separable. The Boolean equation associated with each operator is the representative equation of the corresponding unate Boolean function class. These Boolean equations characterize the transition to DATA function for each threshold operator.

4.3 2NCL COMBINATIONAL EXPRESSION

A 2NCL combinational expression is different from the common experience. It must not only express the desired data function. It must also express the NULL function, it must express the completeness criterion with its state-holding behavior, and it must isolate orphan paths. 2NCL combinational expressions are also different in that they can have variables with more than 2 values.

The transition to NULL function is identical for every operator and for any acyclic combination of operators. It is universally inherent in any combinational expression, and it does not need to be explicitly specified. One can specify a combinational expression solely in terms of its data function. However, the completeness of the data function and the isolation of orphan paths is not inherent in any expression and must be explicit in the expression of the data function.

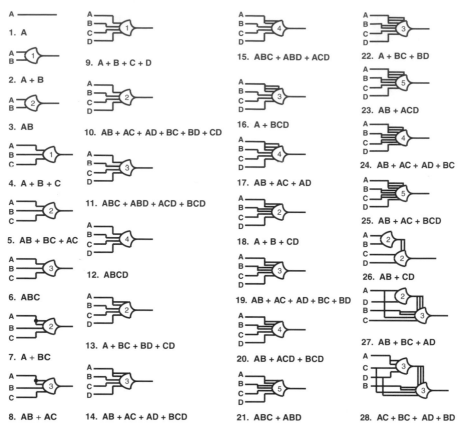

Figure 4.6 The library of 28 threshold operators.

4.3.1 The Two Roles of Boolean Equations

Boolean logic, despite its ultimate expressional shortcomings, is a convenient and commonly understood language of specification. This convenience of specification is used in two ways in the context of 2NCL, which must not get mixed up.

The first usage is the Boolean equation that represents an operator DATA function. This Boolean equation only characterizes the DATA behavior of an operator. This Boolean equation does not fully characterize the behavior of the operator that must include its NULL function and its state-holding behavior. Also the NCL operator does not implement the Boolean function of its associated Boolean DATA function equation.

The second usage of a Boolean equation is as a Boolean function on binary variables. Remember that 2NCL is a subvariable logic whose operators deal directly with values of variables, not with whole variables. Any 2NCL combinational expression on binary variables requires at least 2 operators to output the two paths of a 2 value result variable. A 2NCL operator outputs only one path. Again, a

2NCL operator does not implement a 2NCL combinational expression of its DATA function Boolean equation.

The point is illustrated in Figure 4.7. In Figure 4.7*a*, the A and B of the Boolean equation represents individual values presented to the input of the 2NCL operator. The Boolean equation associated with the operator expresses the transition to DATA function for the operator. The operator is not a 2NCL implementation of the Boolean function. The Boolean function, where A and B represent whole binary variables, is shown in Figure 4.7*b* with the 2NCL implementation of the Boolean function. Operator 2 does not even appear in the 2NCL expression of the Boolean function $F = A + B$.

4.3.2 Combinational Synthesis

Figure 4.8*a* shows a general two-variable function with two input variables X and Y and one output variable Z. A 2NCL variable can have any number of values. Figure 4.8*b* shows a two-variable function with a 2 value input variable, a 3 value input variable, and a 4 value output variable. Each value path of the output variable must be generated by an individual output expression, as shown in Figure 4.8*c*.

For each DATA wavefront each output expression asks whether its output value should transition to DATA given the combination of input values that have transitioned to DATA. While the function as a whole with multi-value variables cannot be characterized by a Boolean equation, each individual output expression, because it deals with the binary DATA-NULL states of individual values during a DATA wavefront, can be characterized with a Boolean equation. It is convenient to specify each output expression as a Boolean sum of products equation and then map each Boolean equation to a 2NCL expression.

There are no signal inversions in a 2NCL combinational expression. With the monotonic behavior of the DATA path, there are no transitions from DATA to NULL during a DATA wavefront. Consequently a Boolean equation characterizing an output expression for a DATA wavefront is strictly unate. Remember that the library of 2NCL operators was constructed to map to all unate Boolean functions of four variables or less. The DATA function unate Boolean equations of the library operators map directly onto the unate Boolean output equations.

The complete behavior of a multi-value variable is characterized by the collective behavior of its set of Boolean output expressions, one equation for each value of the variable. Exactly one output value per variable must transition to DATA for each

a. 2NCL operator 2 and the Boolean equation of its DATA threshold function

b. Boolean function and its 2NCL combinational expression

Figure 4.7 Two roles of Boolean equations.

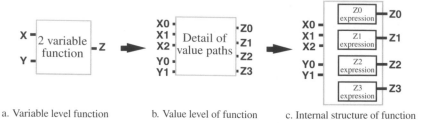

a. Variable level function   b. Value level of function   c. Internal structure of function

Figure 4.8 Mapping a two-variable multivalue function.

DATA wavefront. Since this single transition must indicate completeness of input, each output expression must individually express the completeness of the input variables and must also isolate orphan paths. This can be conveniently accomplished by retaining the sum of products form of the Boolean equations when mapping to 2NCL operators. As will be shown later, optimizing the Boolean equations before mapping can greatly confuse the issues of completeness and orphan paths. Optimization must be performed in terms of 2NCL operators after the mapping.

4.4 EXAMPLE 1: BINARY PLUS TRINARY TO QUATERNARY ADDER

Construction of the 2NCL combinational expression begins with determining the individual Boolean sum of product expressions for each output value. The first example will be the addition of a binary variable to a trinary variable producing a quaternary variable. The structure of the function is identical to Figure 4.8 and requires four output expressions. The function can be characterized by the function map in Figure 4.9 and the Boolean output equations can be read directly off the map. For instance, $Z2 = X2Y0 + X1Y1$. The output equations are then mapped to generic equations that map directly to 2NCL operators. Boolean logic provides a convenient intermediate specification language.

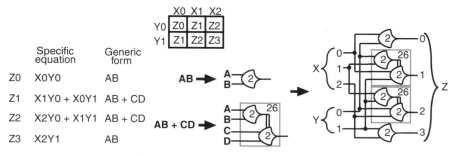

Figure 4.9 2NCL mixed radix adder.

The four output expressions in this case can be implemented with single operators. Function AB is implemented by a 2 of 2 operator. Z0 and Z3 requires one DATA from each input variable so the completeness criterion is fulfilled. Function AB + CD is implemented by operator 26. Z1 and Z2 also requires one DATA from each input variable fulfilling the completeness criterion. The resulting 2NCL expression of the multi-value function in the right of Figure 4.9 is implemented with four operators. There is only one rank of logic, the expression as a whole fulfills the completeness criterion, and no ineffective path transitions can get past the one rank of operators. So orphans are isolated to path segments within the expression.

4.5 EXAMPLE 2: LOGIC UNIT

The next example is a logic unit that will perform an AND, OR, or XOR on two binary variables. The function has two 2 value input variables, one 3 value input control variable with values A (AND), O (OR), and X (XOR), and one 2 value output variable. The function is shown in Figure 4.10. The equations that read off the function map are somewhat more complex than those of the previous example. They do not map directly to single operators and have to be resolved in stages. There are two ways to partition an expression to resolve in stages. The products can be partitioned or the sums can be partitioned. Product partitioning will be used in this instance. In product partitioning two input variables are fully resolved, producing an internal variable. Then the next input variable is resolved in relation to the internal variable producing a second interval variable, then a next input variable is resolved in relation to the second internal variable, and so on until the output variable is produced. This is a straightforward way to proceed, and it ensures progressive logically determined variable boundaries through the combinational expression. Each logically determined internal variable boundary can express completeness and isolate orphan paths.

In the 2NCL expression of Figure 4.10 the first rank of 2 of 2 operators resolves the X and Y inputs into a 4 value variable with the values s, t, u, and v. This first rank

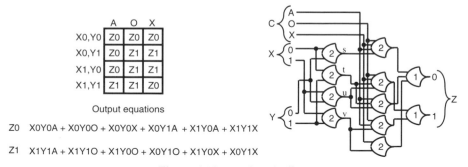

	A	O	X
X0,Y0	Z0	Z0	Z0
X0,Y1	Z0	Z1	Z1
X1,Y0	Z0	Z1	Z1
X1,Y1	Z1	Z1	Z0

Output equations

Z0   X0Y0A + X0Y0O + X0Y0X + X0Y1A + X1Y0A + X1Y1X

Z1   X1Y1A + X1Y1O + X1Y0O + X0Y1O + X1Y0X + X0Y1X

Figure 4.10   Logic unit slice.

of operators fulfills the completeness criterion on the X and Y input variables and isolates the orphan paths of X and Y. This internal variable becomes an input variable to the second rank of operators.

The second rank of operators will resolve the control input variable with the 3 values A, O, and X with the 4 value internal variable. The new output equations are shown in Figure 4.11. Each equation is still a large sum of products, so the equations have to be resolved in stages. In this instance sum partitioning is applied to the sum terms. Terms or groups of terms (subsums) that map to 2NCL operators are expressed individually and then the subsums are summed. Figure 4.11 shows the chosen sum partitioning. Each subsum can be expressed with one of two possible operators.

The equation $sA + sO + sX$ maps to the generic form $AB + AC + AD$. The operator in the library that implements this specific equation is operator 17, a threshold 4 operator with a weight of three on the A input. This ensures that if B, C, and D are independent, the operator will not respond to any combinations of B, C, or D without A.

In this case, however, B, C, and D (A, O, X), being values of a single variable, are mutually exclusive and combinations of them cannot occur. The 2 of 4 operator 10 implements the generic equation $AB + AC + AD + BC + BD + CD$. Since B, C, and D are mutually exclusive the terms BC, BD, and CD will never be presented to the operator making its behavior in the expression identical to operator 17. So either operator can be used in the expression. Which operator is used will depend on the technology and implementation of the library. With a switching technology such as CMOS, operator 17 will recognize fewer combinations than operator 10 and will therefore be a simpler operator with fewer transistors. In a threshold-based technology operator 10 has simpler threshold relationships than operator 17 and could be the more efficient operator. In drawing these example expressions the choice is often purely aesthetic.

The other subsums have similar options. The output equation $uO + tO + uX + tX$ with the generic equation $AC + BC + AD + BD$ is directly resolved by operator 28. Again, the mutual exclusivity of the inputs allows 2 of 4 operator 10 to be used with the generic equation $AB + AC + AD + BC + BD + CD$. In this case A and B are mutually exclusive and C and D are mutually exclusive, masking out terms AB and CD.

The output subsums $tA + uA$ and $vA + vO$ with the generic equation $AB + AC$ is directly resolved by operator 8 a threshold 3 operator with a weight of 2 on input A. The 2 of 3 operator 5 resolves $AB + BC + AC$, but for both subsums B and C are mutually exclusive and the BC term is masked.

The resulting 2NCL expressions are shown Figure 4.12b with the direct map operators and in Figure 4.12a with the operators taking advantage of the mutual exclusivity relationships. The result of the sum partitioned rank of operators is a 5 value internal variable, whose values are then collected through the threshold 1 operators into the 2 value output variable. Each operator of the second rank requires a value from the internal variable and a value from the control variable, fulfilling the completeness criterion and isolating orphan paths from the two variables.

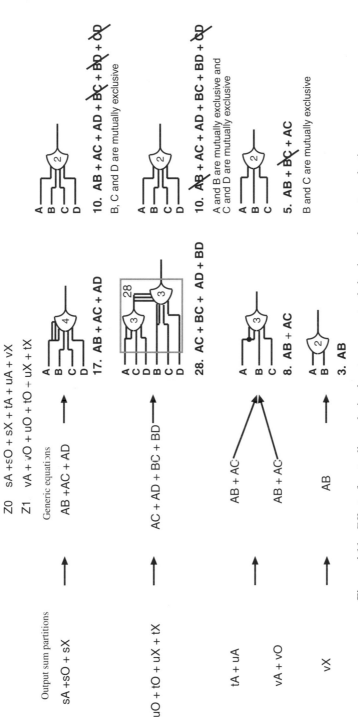

Figure 4.11 Effect of mutually exclusive inputs on operator behavior and operator selection.

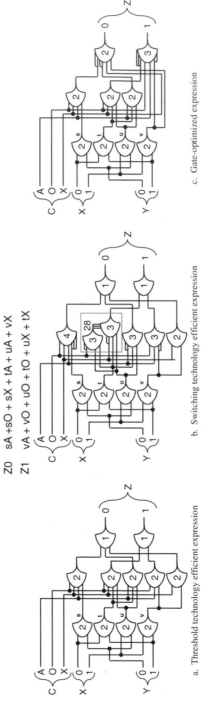

$Z0 \quad sA + sO + sX + tA + uA + vX$

$Z1 \quad vA + vO + tO + uX + tX$

a. Threshold technology efficient expression

b. Switching technology efficient expression

c. Gate-optimized expression

Figure 4.12 Final 2NCL expressions of logic unit.

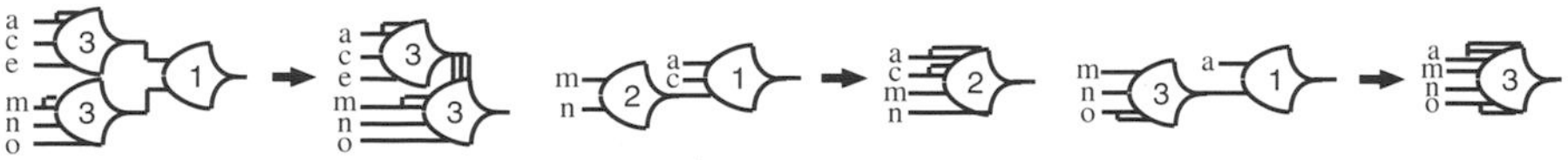

Figure 4.13   Some operator merge optimizations.

There is one more point to be made about the possibility of optimization after mapping the 2NCL operators from the Boolean equations. Two operators can be eliminated by merging the lowest operators of the second rank of Figure 4.12b with the threshold 1 operators shown in as Figure 4.12c. Presenting the output of one operator to another operator with a weight equal to the threshold of the other operator is identical to combining the operator outputs through a threshold 1 operator. This relationship is illustrated in Figure 4.13. While there are fewer operators in the expression, it is not clear that creating a more complex operator to save a threshold 1 operator would be superior in any technology. The threshold 1 operators are very cheap, and the more complex operators are more expensive.

Structuring 2NCL combinational expressions in terms of progressively generated internal variables is a straightforward way to ensure the fulfillment of completion properties and orphan isolating properties of the combinational expression as a whole. Each internal variable can form a completeness boundary that both propagates progressive completeness through the expression and isolates orphan paths.

4.6   EXAMPLE 3: MINTERM CONSTRUCTION

Another approach is to construct the whole minterm of the input variables, which is pure sum partitioning. A rank of M of M operators, with M equal to the number of input variables, individually recognizes each product term. Each operator requires an input from each input variable, fulfilling the completeness criterion and isolating orphan paths. Only one operator will assert DATA for each input DATA wavefront, producing a single large internal variable whose values are collected directly to the values of the output variable or variables. The minterm approach is always safe, reliable, and robust. The construction is simple, completeness is fulfilled, and orphans are isolated. Figure 4.14 shows the minterm expression for the logic unit of example 2 as well as a binary full adder and the binary-trinary-quaternary adder of example 1.

The minterm form may actually be optimal in some cases if multiple variables can be tapped off a single minterm. Consider the binary half adder with two input variables and two output variables. Figure 4.15 shows the function maps and the output equations for the binary half adder. There are two equations for the sum and two equations for the carry. It is clear from inspection that the sum and carry use the same set of product terms. Two approaches can be taken to generating the output expressions. The first is to use the minterm of the common terms to generate a four value internal variable from which values can be collected for both output

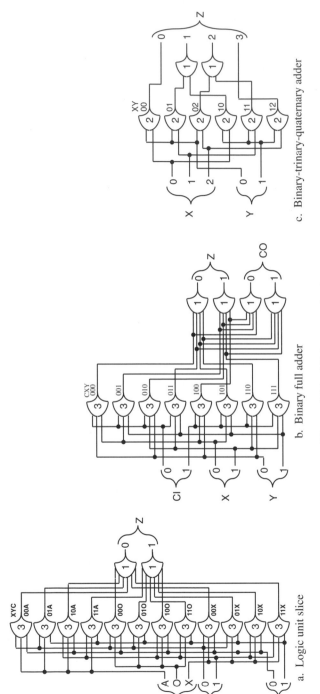

Figure 4.14 Minterm constructs.

a. Logic unit slice

b. Binary full adder

c. Binary-trinary-quaternary adder

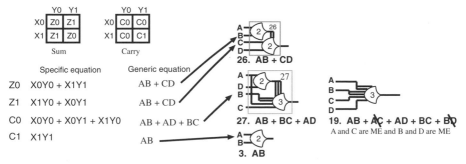

Figure 4.15   Binary half-adder function map and output equations.

variables shown in Figure 4.16*a*. The second is to generate the output expressions directly from the equations as shown in Figure 4.16*b*, *c*, and *d*. Since the output equations map directly to single operators no partitioning of the equations is needed.

The $Z0$ and $Z1$ output values of the sum variable have the same generic equation and can be implemented from two identical operators shown in Figure 4.16*b*. The $C0$ value of the carry variable has a generic equation that is implemented by operator 27. The $C1$ value is just AB a 2 of 2 operator 3. The two operator implementation of the carry output variable is shown in Figure 4.16*c*.

Mutual exclusivity allows an alternative implementation of the carryout expression. Operator 19 resolves the equation $AB + AC + AD + BC + BD$. Since

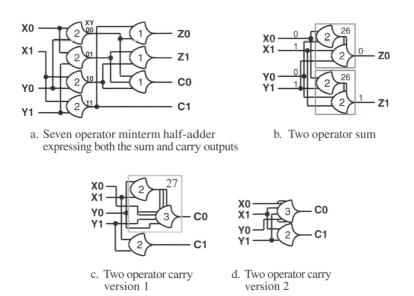

a. Seven operator minterm half-adder expressing both the sum and carry outputs

b. Two operator sum

c. Two operator carry version 1

d. Two operator carry version 2

Figure 4.16   Half-adder minterm expression and directly mapped expressions.

A and C are mutually exclusive and B and D are mutually exclusive, operator 19 can be substituted for operator 27 in the output expression, as in Figure 4.16*d*.

The shared minterm expression of Figure 4.16*a* has seven operators. The half-adder expression generated directly from the four output equations has four operators. However, these four operators are more complex than the seven operators. The shared minterm version of the half adder could very well be the more efficient version.

2NCL minterm construction is similar to Delay Insensitive Minterm Synthesis (DIMS), which is expressed in terms of C-elements and OR functions and 2 value variables [53].

4.7 EXAMPLE 4: A BINARY CLIPPER

An 8 bit to 6 bit 2s complement clipper is shown in Figure 4.17. Bit 7 is the sign bit and is passed through as the sign bit of the 6 bit number. Bits 5, 6, and 7 determine how the low-order bits will be treated: whether they will be all forced to one, all forced to zero, or passed as is.

4.7.1 The Clipper Control Function

The control signals generated from the first three high-order bits express the three meanings; force to 1, force to 0, or pass as is. In the Boolean expression these three control meanings are encoded in two binary variables and applied to the low-order bits in two stages of logic. If one considers the function map in Figure 4.18, it is clear that the control variable is inherently a 3 value variable. Reading the output equations from the map and partitioning the sums yields only two forms of generic equations: ABC, which is implemented with a 3 of 3 operator, and ABC + ABD, which has two possible operators because C and D (60 and 61) are mutually exclusive. The resulting output expressions for the clipper control function are shown in Figure 4.19*a* and *b*. The combinational expression is shown in Figure 4.19*c*.

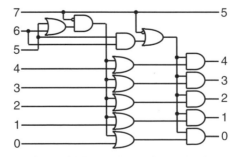

Figure 4.17 Boolean clipper circuit.

	50,60	50,61	51,60	51,61
70	pass input	force ones	force ones	force ones
71	force zeros	force zeros	force zeros	pass input

	Specific equations	Generic equations
pass input	706050 + 716151	ABC + ABC
force zeros	715160 + 715060 + 715061	ABC + (ABC + ABD)
force ones	705160 + 705161 + 705061	(ABC + ABD) + ABC

A
B
C

6. ABC

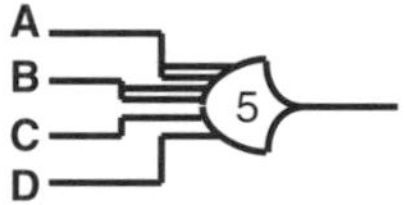

A
B
C
D

21. ABC + ABD

A
B
C
D

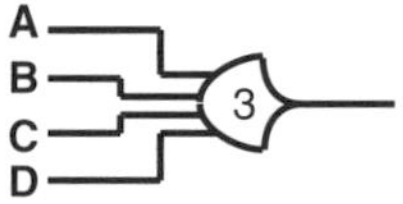

11. ABC + ABD + ACD + BCD

C and D are mutually exclusive

Figure 4.18   Clipper control function.

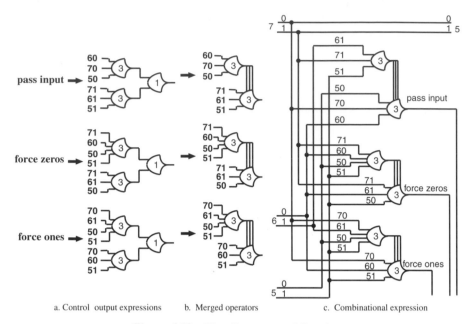

a. Control output expressions   b. Merged operators   c. Combinational expression

Figure 4.19   The clipper control function.

The control can also be implemented by partitioning the product. Bits 5 and 6 are resolved to an internal variable, and then bit 7 is resolved with the internal variable.

Figure 4.20 shows the output equations with the internal variable and their mapping to operators. The resulting expression is shown in Figure 4.20*a*. Figure 4.20*b* is an optimization that combines two operators in the first rank into a single operator, which also simplifies operators in the next rank of logic by lessening their inputs.

4.7.2   The Danger of Minimizing the Boolean Output Equations

Properties of logically determined completeness boundaries derive from the sum of products form of the Boolean output equations. If one sticks with product partitioning and with sum partitioning of the sum of product Boolean output equations, fulfillment of the completeness criterion and isolation of the orphans will inherently follow. Because of the completeness relationships there are no 'don't cares' in NCL expressions. If one performs a standard Boolean minimization of the output equations that removes Boolean 'don't cares' before mapping to 2NCL operators, neither completeness nor orphan isolation inherently follow, and these properties must be explicitly restored.

Figure 4.21 illustrates the preminimization of the output equations for the clipper control. The 'force zeros' and 'force ones' equations can be minimized in terms of Boolean logic to eliminate the 51 and 61 values and the 50 and 60 values, respectively, from the force zeros and force ones equations. The direct mapping of the resulting generic equation results in a 2NCL expression that does not fulfill the completeness

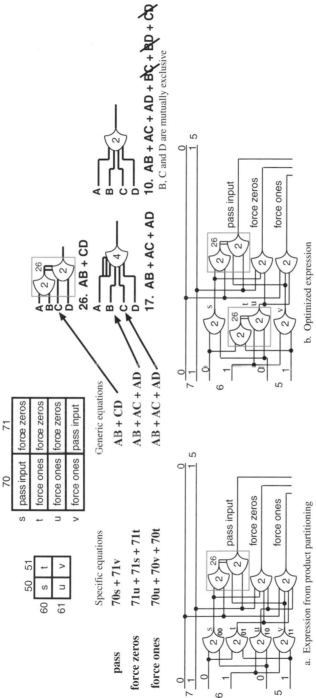

Figure 4.20 Clipper control implemented with product partitioning.

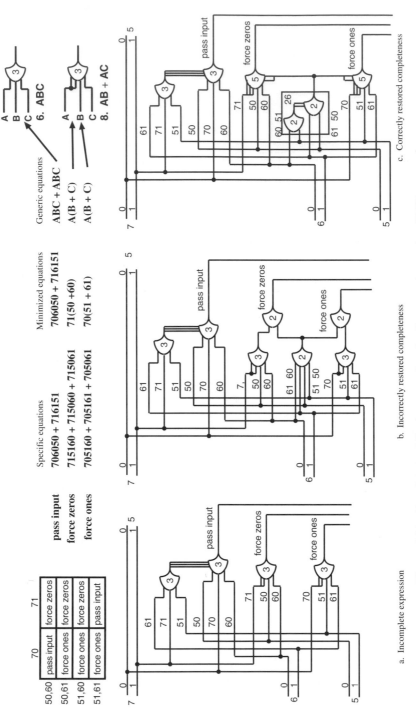

Figure 4.21 Preminimized Boolean equations leading to an incomplete 2NCL expression.

criterion. 'Force zeros' can be asserted with an incomplete input with just 71 and one of 50 or 60.

Completeness must be explicitly restored. One might attempt to restore completeness with the expression of Figure 4.21*b*. While this restores completeness for the 'force zeros' and 'force ones' expressions, the 2 of 4 operator will always achieve complete input and assert an output DATA value producing an orphan through the operator when the 'pass input' result is asserted. The expression of Figure 4.21*c* restores completeness to 'force zeros' and 'force ones' and does not produce an orphan.

An error can be made on a function as simple as the Boolean AND function. Figure 4.22 shows the mapping with preminimized output equations leading to a 2NCL expression that does not fulfill completeness. The output can assert a DATA value when only one input is DATA, as shown with the wide path through the expression.

When one enters the realm of incomplete mappings that must be restored to completeness, one enters a very subtle and uncertain territory [25]. There is no developed methodology for restoring and verifying completeness and of verifying the isolation of orphans for a restored expression. Figure 4.21*b* was long believed by a human mind to be complete and orphan isolating. It was sometime before the presence of the orphan was realized. Figure 4.21*c* was not discovered by a human mind at all but by an experimental program derived from an experimental asynchronous design tool (Petrify [7]). In short, the detection of incompleteness and the restoration of completeness cannot rely on human mentality, and the automated tools are not available.

Although the five operator expression of Figure 4.21*c* for the clipper control is more efficient than the six operator expressions of Figure 4.19 and Figure 4.20, it is most convenient, currently, to map to the 2NCL expression from the sum of products Boolean output equations. This inherently maintains completeness and orphan path isolation in the 2NCL expression. It is then easier to optimize in terms of 2NCL operators while maintaining completeness relationships and orphan path isolation. It is not clear that it will ever be advantageous to minimize the Boolean functions, perform the mapping, and then try to restore completeness and orphan isolation.

4.7.3   The Clipper Data Function

With a 3 value variable as control, the function for each bit 0 through 4 has as inputs one 3 value variable and one 2 value variable, and as output one 2 value variable. The map for this function with the output equations is shown in Figure 4.23.

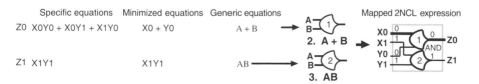

Figure 4.22   Preminimization leading to an incomplete implementation of AND.

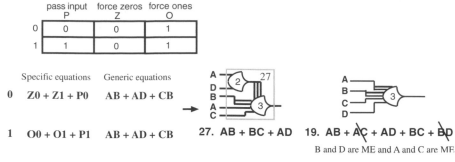

	pass input P	force zeros Z	force ones O
0	0	0	1
1	1	0	1

	Specific equations	Generic equations
0	Z0 + Z1 + P0	AB + AD + CB
1	O0 + O1 + P1	AB + AD + CB

27. **AB + BC + AD**   19. **AB + AC + AD + BC + BD**

B and D are ME and A and C are ME

Figure 4.23   Clipper data path function map.

Figure 4.24 shows an output expression for each bit using operator 19 and a 2NCL expression of the complete clipper.

4.8   EXAMPLE 5: A CODE DETECTOR

Example five shown in Figure 4.25 is the combinational function of a state machine to detect the sequence 0010111 in a sequential stream of binary variables. Figure 4.25a shows the state machine diagram, and Figure 4.25b shows the function map of the combinational expression that resolves the current state with the next input variable to determine the next state. Figure 4.25c shows the output equations derived from the function map. The input of the function is one 2 value variable and one 7 value variable (the current state). The output is one 2 value variable (detect, not detect) and one 7 value variable (the next state). All states output a not detect until state 6 sees a 1 that outputs a detect and returns to state 0 to begin searching again.

The equations for each output value yield two generic equations: AB, which is a 2 of 2 operator, and AB + AC + AD, which is operator 17 and which can also map to operator 10 because of the mutual exclusivity of B, C, and D. Operator 10 is chosen to draw the example expression. The full combinational expression is shown in Figure 4.26. The complete state machine will be presented later, in Section 7.1.3 of Chapter 7.

4.9   COMPLETENESS SUFFICIENCY

Completeness of the output variables must indicate completeness of the input variables and completeness of propagation. If there is only one output variable, then this single variable must fulfill the completeness criterion. If there are two or more variables, their collective completeness must fulfill the completeness criterion. The carry output of the full adder, for instance, does not fulfill the completeness criterion

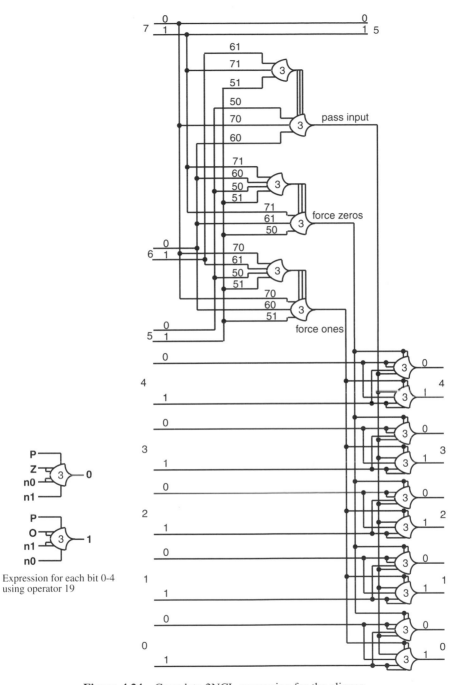

Figure 4.24   Complete 2NCL expression for the clipper.

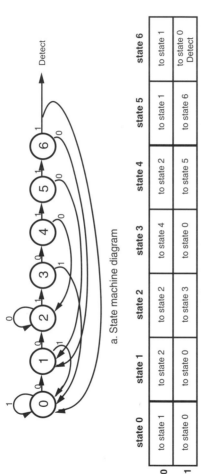

a. State machine diagram

Detect

b. Next state function map

	state 0	state 1	state 2	state 3	state 4	state 5	state 6
0	to state 1	to state 2	to state 2	to state 4	to state 2	to state 1	to state 1
1	to state 0	to state 0	to state 3	to state 0	to state 5	to state 6	to state 0 Detect

	Specific equations	Generic equations
state 0	(1S0 + 1S1 + 1S3) + 1S6	(AB + AC + AD) + AB
state 1	0S0 + 0S5 + 0S6	AB + AC + AD
state 2	0S1 + 0S2 + 0S4	AB + AC + AD
state 3	1S2	AB
state 4	0S3	AB
state 5	1S4	AB
state 6	1S5	AB
detect	1S6	
no detect	everything else	

17. $AB + AC + AD$

10. $AB + AC + AD + \cancel{BC} + \cancel{BD} + \cancel{CD}$

B, C and D are ME

3. AB

c. Output equations and mappings

Figure 4.25 Function map of combinational control function of a state machine.

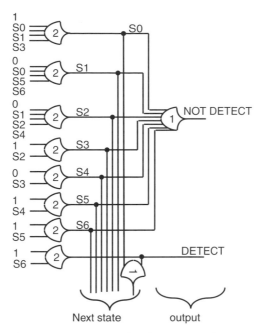

Figure 4.26 Combinational expression for the code detector state machine.

in that the carryout variable can assert DATA with any two of the three inputs being DATA, as shown in Figure 4.27a. But the SUM output variable does require all three inputs to become DATA before it transitions to DATA fulfilling the completeness criterion. The sum variable does not by itself indicate the completeness of propagation. Completion of the sum variable together with completion of the carry variable do fulfill the completeness of propagation as well as completeness of input. Combining the two completions as shown in Figure 4.27b is sufficient to fulfill the completeness criterion for the full-adder expression.

The completeness criterion can be fulfilled even if no single output variable indicates the completion of the input variables. Take an expression with three input variables A, B, and C and two output variables X and Y. Assume that the complete transition of X indicates the complete transition and propagation of A and B and that the complete transition of Y indicates the complete transition and propagation of B and C. The combination of the completeness of X and Y indicates the completeness for A, B, and C.

4.10 GREATER COMBINATIONAL COMPOSITION

Combinational expressions that express completeness and orphan isolation and are bounded by logically determined variables can be combined by connecting variables into greater combinational expressions. The connected variables become internal

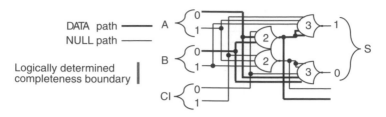

a. Assertion of carry without complete input

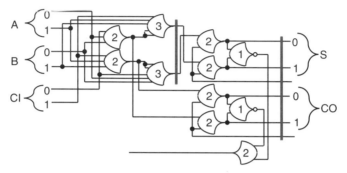

b. Sufficient completeness for full adder

Figure 4.27 Sufficient completeness of full adder.

logically determined variable boundaries of the greater expression, which will itself express completeness and orphan isolation as a whole.

4.10.1 Composition of Combinational Expressions

Figure 4.28 shows a full-adder combinational expression composed of two half-adder combinational expressions and an OR combinational expression. Each subexpression fulfills the completeness criterion and outputs a variable that becomes an internal variable of the greater combinational expression forming internal completeness boundaries in the greater expression.

Product partitioning (see Section 4.5) is essentially decomposing a greater combinational expression into completeness fulfilling subexpressions with logically determined variable boundaries. These boundaries are then recomposed into the greater combinational expression.

4.10.2 The 2NCL Ripple Carry Adder

Combinational expressions with an output variable that does not form a completeness boundary may still be composable if the composition relationship itself leads to the fulfillment of completion. Composing a ripple carry adder from full adders provides an example. The full adders compose through the carry variable that by

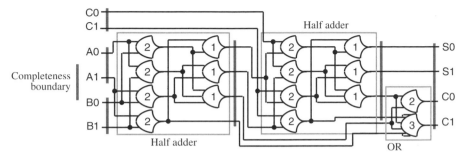

Figure 4.28   Full adder composed of half adders and an OR.

itself does not fulfill the completeness criterion and does not present a completeness boundary. The completeness of each adder stage is expressed by its sum output, which includes the carry from the previous stage. The completeness of all the input of the ripple carry adder is established with the completeness of all the sum variables. The completeness of the propagation of each carry variable, except the last, is established by the sum variable of each following full adder, as shown in Figure 4.29. S0 indicates the completeness of transition and propagation of Cl, A0, and B0. S1 indicates the completeness of transition and propagation of A1 and B1 and carryout from the S0 adder stage. Each Sn indicates the completion of transition and propagation of the Sn − 1 carryout. The completion of propagation of the last carryout is explicitly determined and combined with all the sum completions. While no single output indicates the completeness of all the inputs, the combination of all the sum completions and the last carry indicates the completion of transition and propagation of all the inputs for the whole ripple carry adder combinational expression.

4.11   DIRECTLY MAPPING BOOLEAN COMBINATIONAL EXPRESSIONS

Boolean function expressions can be mapped directly into 2NCL combinational expressions by simply substituting equivalent expressions. While it involves inefficiencies of expression and limits expressions to 2 value variables, this method is attractive because it allows the utilization of standard design tools and methodologies and provides an easily accessible and immediate entry to logically determined system design.

4.11.1   Mapping 2 Variable Boolean Functions

The 2 NCL expressions mapping to 2 variable Boolean functions were introduced in Section 2.3.3 of Chapter 2. They are derived from function maps in Figure 4.15 and Figure 4.16 for the synthesis of the half adder, which is a Boolean exclusive OR for

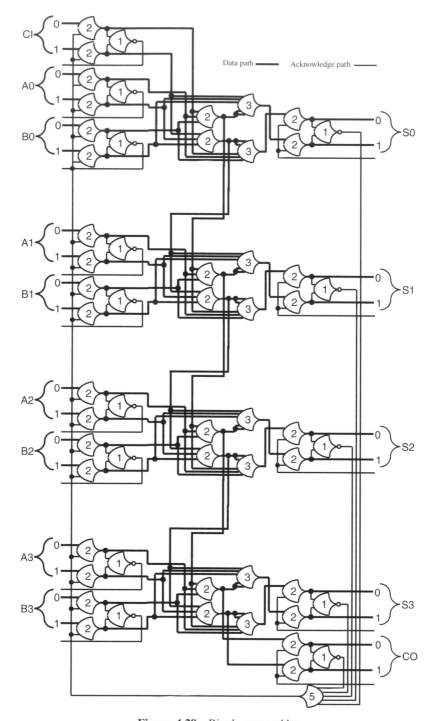

Figure 4.29 Ripple carry adder.

the sum function and a Boolean AND for the carry function. The 2NCL expression for the Boolean OR is just a remapping of the expression for the AND function. Because of the mutual exclusivity of inputs there are two alternate expressions for the AND and OR functions. The set of expressions is shown in Figure 4.30. Any acyclic combination of these expressions will be a logically determined combinational expression. Each expression fulfills the completeness criterion and isolates orphan paths.

Substituting these expressions into any Boolean logic combinational expression will reliably yield a logically determined 2NCL combinational expression. The resulting 2NCL expression fulfills the completeness criterion with progressive completeness boundaries through the expression, and it isolates the orphan paths between internal completeness boundaries.

Figure 4.31a shows the Boolean expression of the 8 bit to 6 bit clipper. Figure 4.31b shows the NCL combinational expression generated by direct Boolean function substitution.

Direct function mapping may be inefficient in terms of transistors, but it is very efficient in terms of fast reliable design. It is simple, convenient, readily accessible, and reliably yields a logically determined expression, and it can be directly supported with standard design tools [28,44].

4.11.2 The Boolean NPN Classes and 2NCL Expressions

The NPN classes include the negation of the output of the function as well as permutation and negation of the inputs. Each NPN class represents two complementary functions from either two complementary NP classes or one NP class that is self-complementary. This structure of complements can be seen in the class table of Figure 4.4. The PN classes are ordered such that complementary functions are neighbors in the table. The NPN classes combine these complimentary PN classes and will be referenced in terms of the Mx numbers at the left of the table. For instance, the PN classes M3 and M4 are complementary and form the NPN class M3/M4. Some NPN

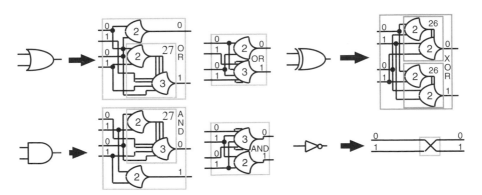

Figure 4.30 2NCL expression for 2 input Boolean functions.

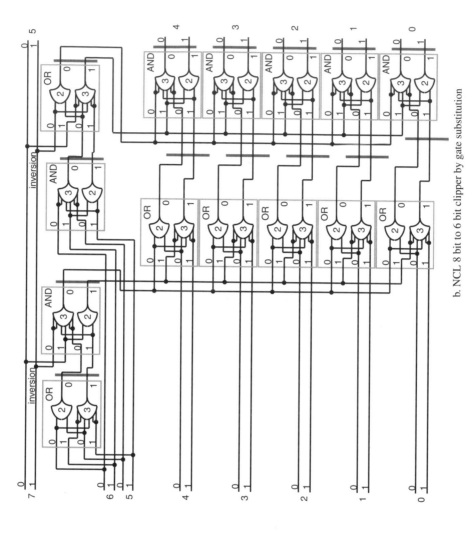

b. NCL 8 bit to 6 bit clipper by gate substitution

a. Boolean logic 8 bit to 6 bit clipper

Figure 4.31 NCL clipper expression by direct substitution.

classes correspond to only one PN class. In this case the PN class, M5, for instance, is self-complementary and provides both complementary functions to the NPN class M5/M5.

A 2NCL expression of a Boolean function must explicitly express the resolution of the 1 output and the 0 output of the Boolean function. To express the completeness criterion the two output expressions must be complementary. These complimentary output expression are the complementary expressions of an NPN class. Consequently the complement of the Boolean function can be expressed by just connecting the 2NCL expression upside down. Since for 2NCL, variable inversion is just a matter routing connections, both the turning upside down and the inverting of the variables is a matter of properly connecting paths.

This is illustrated in Figure 4.32, showing that the OR expression is just the AND expression upside down with paths crossed, and vice versa. Both the Boolean AND and OR functions, members of the NPN class M3/M4, can be implemented with a single 2NCL expression by properly connecting the input and output paths. The significance of this is that any 2NCL expression of a Boolean function that expresses the completeness criterion is inherently a representative expression of an NPN class, and all the Boolean functions of the NPN class can be expressed simply by properly connecting the representative 2NCL expression.

There is a direct correspondence between the NPN Boolean function classes and 2NCL expressions. Just two 2NCL expressions, the AND/OR expression and the XOR expression, corresponding to the two NPN classes of two variable Boolean functions M3/M4 and M5/M5 are sufficient to substitute for any two variable Boolean function.

4.11.3 Mapping NPN Classes for Three-variable Boolean Functions

Recall from Table 4.1 that the 256 three variable or less Boolean functions map into 14 NPN classes. One of these classes is a constant, and one is a single variable class, two are two variable classes. This leaves 10 NPN classes of Boolean functions of exactly three variables. This means that 12 NCL expressions derived from the

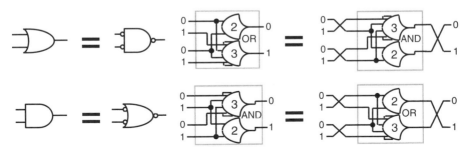

Figure 4.32   NPN function class M3/M4.

NPN classes can directly map to all 256 possible Boolean functions of three or fewer variables.

Figure 4.33 shows the derivation of the 2NCL representative expression for NPN class M10/M11. The 2NCL combinational expression of Figure 4.33*b* will map to 48 of the 218 three variable Boolean functions.

The operators of the clipper of Figure 4.31 can be grouped into three variable Boolean functions of NP class M10 and NP class M11. Each of these can be directly substituted with the representative expression of NPN class M10/M11 shown in Figure 4.33. The entire clipper can be expressed with seven instances of a single expression, as shown in Figure 4.34.

This may not be the most efficient 2NCL implementation of the clipper. It is simply one of the most expedient implementations. The way it is shown looks like it uses 42 operators, though it is really just seven instances of a single library macro. The previous expression of Figure 4.24 used only 16 operators but with 16 library macros.

The 12 NPN representative expressions can be implemented as library macros, and they can be conveniently mapped with current automated design tools onto any Boolean expression in functions of three or less variables reliably forming a logically determined 2NCL combinational expression.

4.12 SUMMARY

Threshold function classification and Boolean function classification have provided a firm conceptual foundation for a 2NCL synthesis methodology. Function

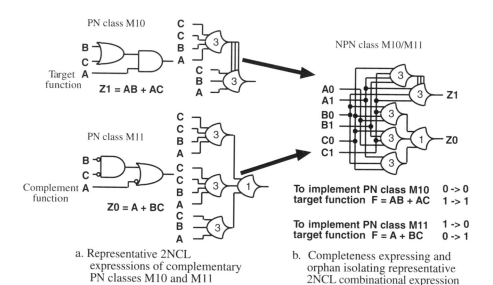

Figure 4.33 Representative 2NCL expression for NPN class M10/M11.

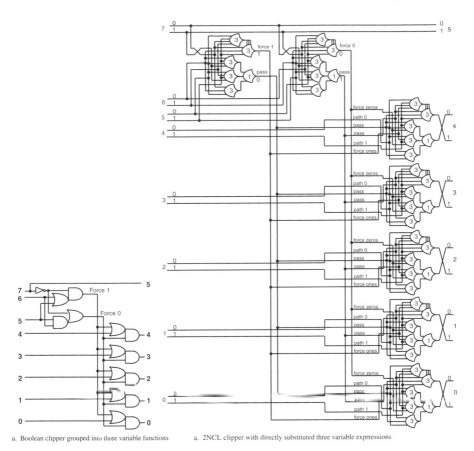

a. Boolean clipper grouped into three variable functions   a. 2NCL clipper with directly substituted three variable exprressions

Figure 4.34 Clipper implemented by direct substitution of three-variable Boolean functions.

classification provides a rationale for the construction of a 2NCL operator library as well as the methodology to synthesize directly from function tables with multi-value variables. It also provides the basis for the direct mapping of Boolean expressions into 2NCL expressions.

4.13   EXERCISES

4.1. Implement the library of operators in terms of S-R flip flops as macros in your available simulation program or FPGA development environment. (See the discussion on playing with 2NCL in Appendix B.)

4.2. Simulate the examples of this chapter.

4.3. Design and simulate five common functions.

4.4. Define a procedure to synthesize 2NCL combinational expressions directly from any multi-value function map.

4.5. Define the set of completeness preserving and orphan isolating optimization transformations on 2NCL operators.

4.6. Define a procedure to apply optimizations to any 2NCL combinational expression.

4.7. Define the library of 12 2NCL representative expressions for the NPN classes, and determine the connection mappings for all 256 Boolean functions.

CHAPTER 5

Cycle Granularity

The cycles of a logically determined system can be any granularity. They can be large or they can be very small. System performance depends on the period it takes a cycle to cycle. Large cycles are slower than small cycles so a critical factor of logically determined system design is the degree to which the size of cycles can be minimized [45,46,47,48]. The cycle period depends on the cycle path shown in Figure 5.1. The cycle path includes the spanning completion detection path, the acknowledge path, the spanning completeness path of the acknowledge signal, and the data path between the spanning acknowledge and the completion detection, which may include a combinational expression. As examples get visually more complex, cycles will be indicated by highlighting the completion/acknowledge path of the cycle, as shown in Figure 5.1.

Smaller cycles can be achieved by partitioning combinational expressions, by partitioning the data path, and by integrating the data path combinational logic and control logic. The smallest possible cycle consists of a single 1 value variable data path with no combinational expression shown in Figure 5.2

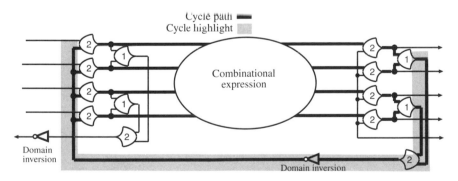

Figure 5.1   Cycle path and cycle highlighting convention.

Logically Determined Design: Clockless System Design with NULL Convention Logic^{TM}, by Karl M. Fant
ISBN 0-471-68478-3   Copyright © 2005 John Wiley & Sons, Inc.

Figure 5.2   Minimum possible cycle.

5.1   PARTITIONING COMBINATIONAL EXPRESSIONS

A combinational expression can be partitioned with the familiar pipeline slicing into data path wide cycles. It can also be partitioned into even smaller units that correspond to the generation of each variable in the combinational expression.

5.1.1   Pipeline Partitioning the Combinational Expression

The expression of Figure 5.3 will be used as the first discussion example. It is a single cycle with a combinational expression that is a binary full adder expressed as two half adders and an OR function. The first thing to understand is that any rank of combinational logic can be turned into a registration stage.

A spanning completeness path is formed by presenting an acknowledge to each operator as a necessary input. Regardless of the function of the operator, one more necessary input means the threshold of the operator must be increased by one. An M of N operator becomes an M + 1 of N + 1 operator. The completeness path must span the data path. If there are paths that bypass the rank of logic, the rank of logic must be extended through these paths by adding 2 of 2 operators and the acknowledge. Spanning completion detection is then added after the spanning completeness path, and one has a registration stage.

This process is illustrated in Figure 5.4. Each rank of logic in the combinational expression has been turned into a registration stage. Where there was one very long

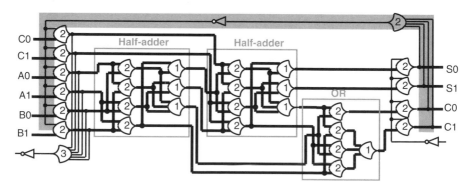

Figure 5.3   Single cycle with an example combinational expression.

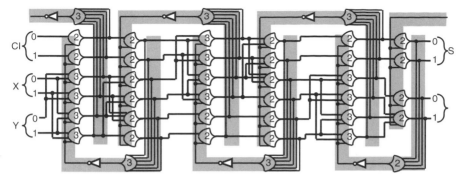

Figure 5.4 Example with all logic stages converted to registration stages.

cycle there are now five smaller cycles with faster periods, and the throughput for the expression is increased.

One can also view this partitioning in terms of merging the combinational logic into regulator ranks. Figure 5.5 shows the example combinational expression with the 2 of 2 ranks of logic turned into registration stages. The remaining ranks of threshold 1 operators can be merged into the regulator ranks.

The expression of Figure 5.6a can be merged into the expression of Figure 5.6b. If the inputs B, C, and D are mutually exclusive, which they are in this case because they are values of a minterm variable, then the inputs of the threshold 1 operator can be merged into the operator shown in Figure 5.6c.

The resulting expression of Figure 5.7 has fewer cycles than Figure 5.4 with more combinational logic integrated into each spanning completeness path. The version of Figure 5.7, with fewer operators, may have fewer transistors and the version of Figure 5.4, with simpler faster operators, may have higher throughput.

The reader should notice that in the expressions of both Figure 5.4 and Figure 5.7 there is not a single operator performing solely a data path combinational function.

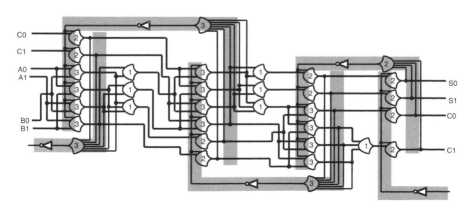

Figure 5.5 Example with 2 of 2 logic stages converted to registration stages.

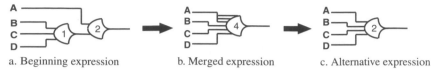

a. Beginning expression   b. Merged expression   c. Alternative expression

Figure 5.6   Merged operators.

In both instances the entire combinational expression has been merged into spanning completeness paths of registration stages. The expression of flow control and the expression of data path combinational functionality are fully merged.

It may not always make sense to integrate combinational logic and control this thoroughly. For instance, the full-adder expression of Figure 5.8a has many paths bypassing the first rank of logic and turning that rank of logic into a spanning completeness path is rather expensive in terms operators and in terms of the completion detection. The pipelined version of Figure 5.8b may not perform any better in any sense than its nonpipelined version of Figure 5.8a. Also the merging of the flow control and combinational logic expressions may require complex operators such as the 5 input threshold 4 operators in Figure 5.8b and the 3 of 5 operators of Figure 5.7 that may not be available in the libraries. It might be very useful to extend the library to include all four input operators extended by one necessary input to accommodate merging acknowledges.

5.1.2   Variable Partitioning the Combinational Expression

A combinational expression may be partitionable into even smaller cycles. Each stage of variable generation in the combinational expression can be turned into a registration stage. Figure 5.9 shows the expression of Figure 5.4 further partitioned at the level of individual variables. The partitioning may be guided by the acknowledge rules of cycle construction. Every source registration stage must be

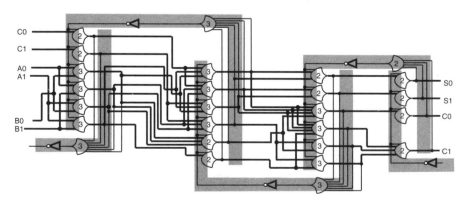

Figure 5.7   Threshold 1 operators merged into regulator ranks.

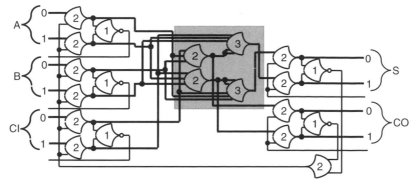

a. Combinational full adder in single cycle stage

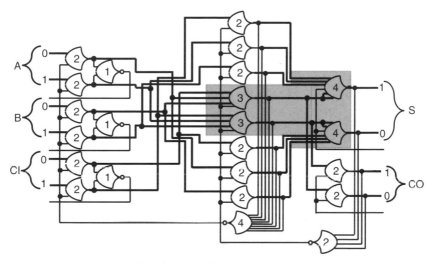

b. Full adder fully pipelined over two cycle stages

Figure 5.8   Optimal full adder nonpipelined and pipelined.

acknowledged by all the destination registration stages it contributed to, and every destination registration stage must acknowledge every source registration stage that contributed to it. The gist is to make the logic asserting each internal variable a registration stage. Essentially, a combinational expression can be woven from cycles, variable by variable.

5.2 PARTITIONING THE DATA PATH

An acknowledge path spanning a wide data path creates a very large cycle. There is no inherent necessity that a data path be spanned by a single cycle with full width

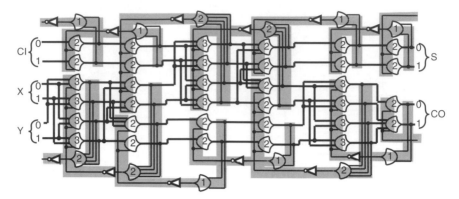

Figure 5.9 Variable partitioned data path with integrated combinational logic.

completion and acknowledge. The urge for full data path completion is a misconception derived from the mathematical abstraction of number as a primitive unit of manipulation and from the experience of controlling the data path with the clock. In the context of this discussion the primitive unit of manipulation is a 2NCL variable. A digit is a 2NCL variable, and a number is an array of digits.

The data path can be partitioned into small cycles, reducing the overall cost of the completion path both in terms of time and resources. If there are no dependency relationships between variables in a data path, as in the case with a transmission pipeline, then each variable in the data path can manage its flow independently. The multi-variable data path cycles of Figure 5.10a can be partitioned into independent single-variable data path cycles of Figure 5.10b. For wide data paths this can save a lot of resources and a lot of time. The independently flowing variables will be re-synchronized when they encounter dependency relationships in the data path such as a combinational expression or a stage that requires full data path synchronization such as an I/O port.

5.3 TWO-DIMENSIONAL PIPELINING: ORTHOGONAL PIPELINING ACROSS A DATA PATH

With a variable partitioned data path, pipelines can be constructed across the data path. This results in orthogonal pipeline structures that will be called two-dimensional or 2D pipelines.

A data path can be partitioned even if there are dependency relationships among the variables such as the adder carry dependence or a data path spanning control variable. These dependency relationship can be pipelined across the data path with 2D pipelining.

The ripple carry binary adder will serve as an introductory example. The carry of a ripple carry adder is a dependency among variables (digits) in the data path. This dependency relationship can be accommodated across the full data path width by

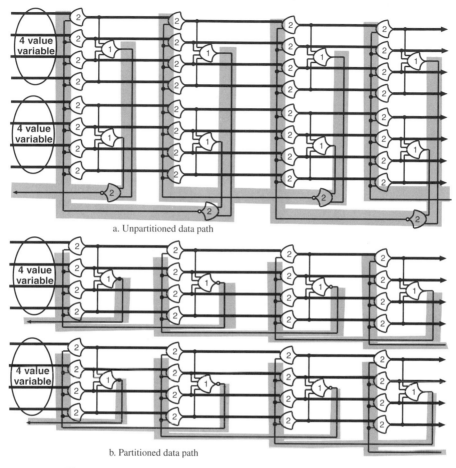

a. Unpartitioned data path

b. Partitioned data path

Figure 5.10   Pipeline with data path unpartitioned and partitioned.

demanding completion of all the output variables as well as the last carryout variable, as shown in Figure 5.11a with the full data path width cycle highlighted.
The dependency relationship can also be accommodated by pipelining the carry
value between the full adder stages from the least significant variable (LSV) to
the most significant variable (MSV) forming a structure of many small cycles
across the data path, as shown in Figure 5.11b with the cycles of the structure highlighted. As data path variables pipeline along the data path through the adder, the
carry variable pipelines across the data path variables. A cycle encompasses each
full adder, sharing completeness paths vertically with the neighboring full adder
cycles and horizontally with the input variable cycles and the output variable cycles.

The pipelining across the data path can be of any granularity. Each orthogonal
cycle can encompass two, four, or more full-adder stages (digits). It is a trade-off
between the cost of completion, the cost of pipelining, and the desired performance

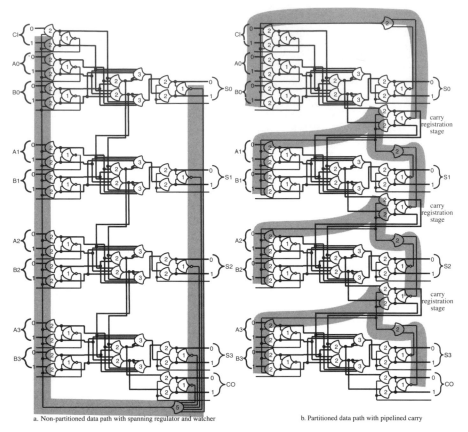

a. Non-partitioned data path with spanning regulator and watcher   b. Partitioned data path with pipelined carry

Figure 5.11 Full-completion adder and 2D pipelined adder with cycle structure highlighted.

parameters. One can also choose the granularity of the variables. While the adder under consideration is a binary adder, it can just as well be a quaternary adder.

Figure 5.12a shows the cycle structure for a full-completion data path, and Figure 5.12b shows the cycle structure for a 2D pipelined data path. Each ellipse

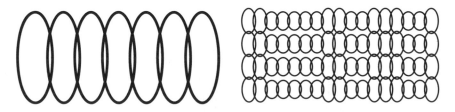

a. Cycle structure of full-completion data path   b. Cycle structure of 2D pipelined data path

Figure 5.12 Cycle structures for full-completion data path and a 2D pipelined data path.

is a cycle, and the overlap of the ellipses represents shared completeness paths (registration stages) between the cycles. All cycles share paths along the data path. In the 2D pipelined data path, some cycles share paths orthogonally to the data path and some do not. There will be cycles for which the variables of a wavefront have no dependency relationships and the individual variables of the wavefront can flow independently along the data path. These include transmission pipelines, buffer stages, logic operations, format conversion operations, and so forth. If there are dependency relationships among the variables of a wavefront, then the cycles must shares paths orthogonally to the data path pipelining dependency relationships across the data path. These include most arithmetic operations, searches such as priority encoding, and control operations that steer wavefronts.

5.4 2D WAVEFRONT BEHAVIOR

The 2D pipelined wavefronts flow diagonally through the data path. When the LSV adder stage has completed its sum variable, there is no logical necessity for it to wait until the MSV sum variable has been completed. The LSV sum variable can proceed to flow through the data path. As soon as the LSV variable stage has completed its carry variable the LSV + 1 adder stage can proceed to complete its sum variable. As the carry variable propagates to the MSV adder stage, sum variables are presented to the data path in succession. The LSV sum variable leads in the data path, and each successive sum variable lags slightly, resulting in a diagonal wavefront of sum variables.

As soon as the LSV sum variable has propagated out, the LSV full-adder stage can receive and propagate the NULL wavefront and then receive and resolve the LSV variable of the next wavefront. There can be multiple wavefronts simultaneously flowing through the ripple carry adder in progressive stages of completion. This produces a much higher throughput than if all the sum variables waited on full data path width completion before proceeding to propagate. This is illustrated in Figure 5.13. The narrow black line in Figure 5.13*a* is the 2D pipelined ripple carry adder. The gray lines are consecutive wavefronts propagating through a 2D pipelined adder. In the full-completion adder, the wide black line of Figure 5.13*b*, the gray lines are the consecutive wavefronts that waited for full width data path completion. The gray line inside the wide black line illustrates the carry propagation

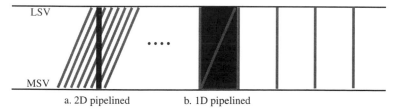

a. 2D pipelined   b. 1D pipelined

Figure 5.13   Wavefront flow for 2D pipelining and full-completion pipelining.

through the full-completion adder and contrasts the two behaviors. The horizontal distance between the lines illustrates that the 2D pipelined wavefronts have a much higher throughput than the 1D pipelined wavefronts.

A full-completion ripple carry adder will deliver average case completion behavior across its sum variables depending on its width, but the 2D pipelined ripple carry adder delivers constant case behavior independent of its width. If the wavefronts presented to the adder are propagating at the appropriate diagonal, then the carry variable of sum Vn will arrive at stage Vn + 1 at the same time than the wavefront variables n + 1 are arriving at the adder stage Vn + 1. The ripple of the carry does not cause any delays. It can be seen from Figure 5.13 that as the full-completion adder gets wider, the carry takes longer and the wavefronts get farther apart. As the 2D pipelined adder gets wider, it just accommodates more concurrent wavefronts, and the distance between wavefronts remains the same. No matter how wide the 2D adder, 8 digits or 1000 digits, its throughput is constant.

5.4.1 Orthogonal Pipelining Direction

The data path can be orthogonally pipelined either from LSV to MSV or from MSV to LSV. Different functions demand pipelining in different directions. Most control variable wavefronts such as MUX and DEMUX select, register file commands, addresses, and so forth can be piped in either direction. The carry dependency of addition flows from LSV to MSV. Other functions, such as compare which typically begins with the sign bit, naturally flow from the MSV to the LSV. Other functions that shuffle variables, such as bit reverse and barrel shift, do not favor diagonal wavefronts in either direction but instead favor a vertical slope. Different dependency relationships result in wavefront flows with conflicting slopes, illustrated in Figure 5.14. LSV leading wavefronts will be called positive slope wavefronts and MSV leading wavefronts will be called negative slope wavefronts. Wavefront slope conflict is the primary issue for 2D pipelining.

5.4.2 Wavefront Conflicts

As mentioned above, if the wavefront is propagating at the right slope, then the arrival of the variables in the wavefront can occur at a variable stage at the same time that the orthogonally propagating dependency arrives and there is no waiting. If the

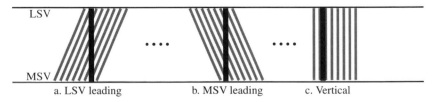

Figure 5.14 Opposing 2D wavefront flows.

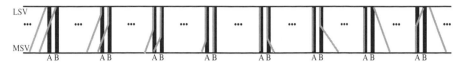

Figure 5.15 Time sequence of wavefront flow through data path functions demanding opposing slope flow.

wavefront is not at the right slope, then there is waiting. If data path functions demand opposing slope flow, there is a lot of waiting. Figure 5.15 shows progressive snapshots of the flow of a wavefront through two stages demanding opposing slope flow. The orthogonal dependency of function A flows from LSV to MSV, so wavefronts flowing through A lead with LSV with a positive slope. The orthogonal dependency of function B flows from MSV to LSV, so wavefronts flowing through B lead with MSV with a negative slope.

The LSV variable emerges from function A, but function B will not be ready for it until its dependency has propagated from its MSV to its LSV. Function B will not begin propagating until the MSV of function A has emerged and is presented to function B. So the LSV result variable of function A wait through the orthogonal propagation of function A and then through the orthogonal propagation of function B before it can propagate through the LSV of function B. The wavefront flattens out against function B and then begins emerging with the opposite flow diagonal. The flow of A is blocked until the LSV of B flows, so the succeeding wavefront flattens out against A waiting for B to propagate. Wavefront flow conflict can detrimentally affect the throughput of a data path. But this can be managed.

5.4.3 Managing Wavefront Flow

The first thing to understand is that no matter how conflicted, congested, or how ragged the wavefront flow becomes along the data path, the system is logically determined and will perform correctly. One must get used to trusting the logic flow. It is very different from trusting state sequencing and timing analysis. Managing the wavefront flow has no impact whatever on the correct logical functioning of the system. The two issues are completely separate.

The key to performance for a 2D pipelined system is managing the slope of wavefront flow. This can be achieved with a combination of buffering and data path function design. The slope of wavefront flow is a function of the forward latency along the path and the forward latency across the data path. Each 2D pipelined cycle will typically have two outputs: a result output that propagates along the data path and a dependency output that propagates across the data path. The forward latency of these two outputs can be different. If they are about the same, it can be said that the wavefront propagates with a slope of 45°. If the across is slower than the along then the slope will be less than 45°. If the across is faster than the along then the slope will be greater than 45°.

The function with the smallest slope will dominate and determine the propagation slope for the system. It is somewhat analogous to the slowest stage of a pipeline. The variable stages of functions with steeper slopes (fast propagation across the data path) will always have to wait on input variables from functions with shallower slopes (slower propagation across the data path).

The throughput of the data path is determined by the distance between wavefronts along the data path. This is determined by how long it takes successive result variables to emerge from a cycle of the data path. So the throughput of the data path is still determined by the slowest cycle of the data path. Another way to say it is that the throughput will be determined by the slower of the horizontal and vertical outputs of the cycle. In this context it is not advantageous for one output to emerge before the other output. There is no advantage, for instance, to compute the carry value faster than the sum value to achieve faster carry propagation. With 2D pipelining, if the carry is produced early, it will just have to wait for the variables on the data path. A fast carry may change the slope of the wavefront, but it will not increase the throughput. The critical factor is the cycled period itself, and there is no throughput advantage to one output occurring faster than another.

5.4.4 Wavefront Slope Buffering

Wavefront slopes can be managed with triangle buffering. The triangle buffer is composed of buffer cycles with no dependency relationships across the data path. Figure 5.16c illustrates a triangle buffer in terms of cycles. The larger number of cycles at the LSV side of the data path form a FIFO that stores and delays the leading variables while the trailing variables catch up. As the buffer crosses the data path, fewer cycles are needed to form smaller FIFOs. Figure 5.16a shows positive slope wavefronts being converted to vertical slope wavefronts through the black

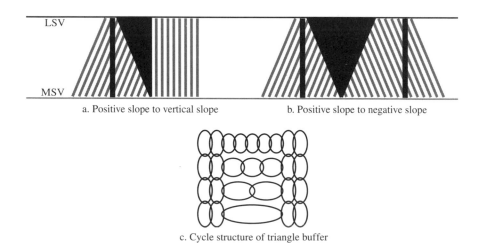

a. Positive slope to vertical slope   b. Positive slope to negative slope

c. Cycle structure of triangle buffer

Figure 5.16   Triangle buffering to manage diagonal wavefront flow.

triangle buffer. The wide end of the triangle buffer has many buffer cycles, and the narrow end no buffer cycles. The vertical wavefronts can be presented to a clocked interface or a barrel shifter, for instance, at the full throughput of the data path. With a thicker triangle buffer in Figure 5.16b, positive slope wavefronts can be converted to negative slope wavefronts with no loss in throughput.

5.4.5 Function Structuring

The next consideration is the structure of the data path functions themselves. A design should take into account the wavefront angle needs of functions as they progress through the data path to minimize the buffering needed to maintain throughput. A worst case structure would be a data path with alternate functions demanding opposing wavefront slopes.

5.5 2D PIPELINED OPERATIONS

The operation stages of the data path can be data operations or they can be control operations. This section provides an example of each.

5.5.1 2D Pipelined Data Path Operations

A data path operation is an integral part of the data path. Wavefronts arrive, flow through, transform, and continue through the data path. The ripple carry adder of Figure 5.11 is an example of a data path function. Three wavefronts, A, B, and carry-in, arrive, flow through the adder and transform, and two wavefronts, the sum and the carryout, continue through the data paths. A binary comparator is another example of data path operation.

A Binary Comparator. A binary comparator can be implemented as a 2D pipelined structure to flow from either the LSV or the MSV. Figure 5.17 shows comparators of each type. The comparator of Figure 5.17a flows from MSV to LSV. The resolution begins with the sign bit. If A is negative and B is positive, then A is LT B. If A is positive and B is negative, then A is GT B. If both signs are equal, then it must be remembered whether the signs were positive or negative to resolve the upstream comparisons. An internal variable with four values—pos, GT, LT, and neg—is pipelined across the digits of the number. At any digit if a GT or LT resolution is determined, then the comparison is resolved and the GT or LT will continue propagating to the LSV, consuming the remaining digits of the number wavefront. The final output at the LSV is a 3 value variable with the EQ, GT, and LT. If the resolution continues to be equal, then the final result is EQ; otherwise, it is GT or LT.

The binary comparator can also be pipelined from LSV to MSV as shown in Figure 5.17b. In this case the highest order bit difference is detected and piped up the digits to the sign bits. The final resolution cannot be determined until the sign bits are encountered in the MSV stage.

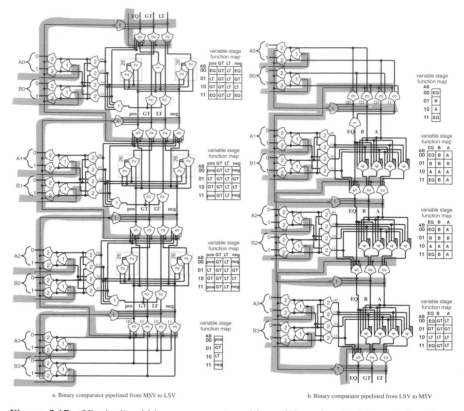

a. Binary comparator pipelined from MSV to LSV

b. Binary comparator pipelined from LSV to MSV

Figure 5.17 2D pipelined binary comparator with partial results piped in both directions.

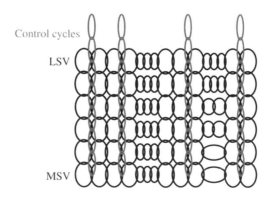

Figure 5.18 Control pipelines in the 2D data path structure.

The two operations are identical. The input wavefronts interact and transform into a single 3 value variable. Each accommodates an opposite signed input wavefront slope. The output wavefront has no slope, since it is a single variable.

The cycle structure of each operation is highlighted. There is a cycle stage for each digit pair of the input numbers. The function map for each digit stage is shown next to the digit stage. The two end cycle stages are different from the middle cycle stages. One end cycle initiates the first partial result that propagates along the digits, and the other end cycle terminates the propagation with the final result of the operation.

5.5.2 2D Pipelined Control Operations

The primary control mechanism of a logically determined system is to steer wavefronts through appropriate data paths. The steering is accomplished through fan-out

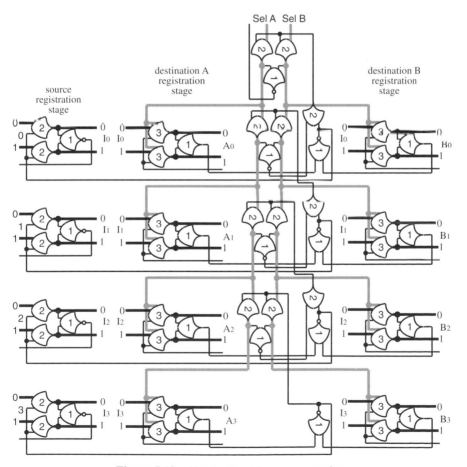

Figure 5.19   2D pipelined fan-out expression.

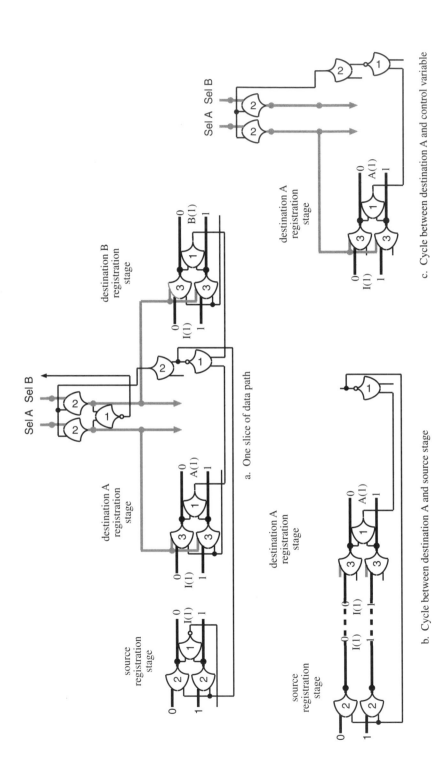

Figure 5.20 Component structures of the 2D fan-in structure.

a. One slice of data path

b. Cycle between destination A and source stage

c. Cycle between destination A and control variable

and fan-in steering structures via a control variable wavefront. In 2D pipelined control the control variable wavefront is pipelined across the data path from the leading variable to the trailing variable. Figure 5.18 shows a data path with the cycles of the gray control variable pipeline entering the data path at the LSV and piping across the data path to the MSV.

2D Pipelined Fan-out Steering. In the fan-out steering expression, a data path wavefront and a control variable wavefront interact to steer the data path wavefront from a source pipeline to one of several destination pipelines (see Section 3.3.2 of Chapter 3). To implement 2D pipelining, the control variable is pipelined across the variables of the data path steering each data path variable in turn. If the data path pipelines from LSV to MSV, the control variable will interact, first with the LSV variable of the data path and then pipe across the data path. It will successively inter-act with the data path variables until the MSV variable is steered, whereupon the control variable is consumed.

A 2D pipelined 1 to 2 fan-out structure is shown in Figure 5.19. At each variable of the data path the control variable enables the destination registration stage for that variable and then flows on to the next variable.

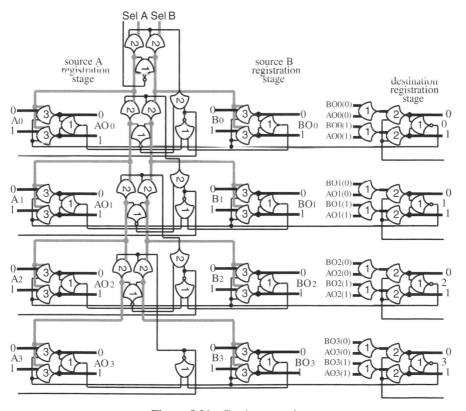

Figure 5.21   Fan-in expression.

Figure 5.20*a* shows one slice of the data path controlled by a control variable pipeline stage. Figure 5.20*b* shows the cycle between the source registration stage and the destination stage A when A is selected. Figure 5.20*c* shows the cycle between the control variable registration stage and the destination stage A when A is selected.

Fan-in. The fan-in expression is the complement of the fan-out expression. The control variable steers multiple source cycles into one destination cycle (see Section 3.3.3 of Chapter 3). The 2 to 1 fan-in expression with the control variable pipeline is shown in Figure 5.21. The component cycle structures of the fan-in structure are compliments of the fan-out components of Figure 5.20.

5.6 SUMMARY

An entire system can be composed of very finely grained cycles. 2D pipelining provides high-throughput behavior for arbitrarily wide data paths. All functions of a system can be 2D pipelined, even the memory. And 2D pipelined wavefront behavior can be managed with triangular buffers. 2D pipelining lends itself particularly to digital signal processing, which can be highly pipelined [43,11,49,26].

5.7 EXERCISES

5.1. Design and simulate 2D pipelined structures for other system components.
   ALU
   Shifter
   Multiplier
   Convolver
   Instruction decoder
   Register file
   Memory wrapper
   Are there any components not amenable to 2D pipelining?

5.2. In a 2D control structure the control value allows a data path wavefront to propagate. The completion of this wavefront implies the control value. Show how this fact can be used to implicitly pipeline the control value avoiding the explicit pipeline of Figure 5.19.

5.3. Define a procedure to synthesize combinational expressions in terms of variable size cycles.

5.4. Define a procedure to synthesize pipeline structure with variable cycle granularity.

5.5. Define a procedure to discover optimal cycle granularity.

CHAPTER 6

Memory Elements

The flow of a wavefront in a logically determined system cannot always be expressed directly in terms of physical path structure. The logical flow path of a wavefront can be directed temporarily through a physical wayside rest until its assigned rendezvous is imminent. Memory elements are these temporary wayside rests for wavefronts in their flow toward resolution.

6.1 THE RING REGISTER

The ring provides a restorative feedback mechanism for wavefront flow that can serve as a wavefront storage register. A wavefront can flow out of the ring and back into the ring, or a new wavefront can flow into the ring replacing the current wavefront. Construction of a ring register begins with the ring of at least three cycles, as shown in Figure 6.1a. There must be an output path from the ring, so another cycle is added that shares a path with one of the ring cycles. This adds the fan-out structure of Figure 6.1b to the ring. This fan-out is not conditional. The wavefront flows from registration stage O to both fan-out destinations. So registration stage O must be acknowledged by both destination registration stages. With this structure a DATA wavefront initialized in the ring will flow to registration stage S, and it will be blocked because registration stage O will not have a request for DATA until the destination registration stage requests DATA. When the destination registration stage requests a data wavefront, the stored DATA wavefront will flow through registration stage O to the destination registration stage and back through the ring to registration stage S. At stage S it will be blocked until the destination registration stage requests DATA again. The cycle including registration stage S and data path S will be called the storage cycle of the ring. Figure 6.1b forms a memory element for a constant wavefront that can be read on demand indefinitely.

To be able to write a different wavefront into the ring, an alternate source for the ring is needed. This source is provided by the fan-in structure of Figure 6.1c. The wavefront that flows to the storage cycle of the ring is either the wavefront already

Logically Determined Design: Clockless System Design with NULL Convention Logic^{TM}, by Karl M. Fant
ISBN 0-471-68478-3   Copyright © 2005 John Wiley & Sons, Inc.

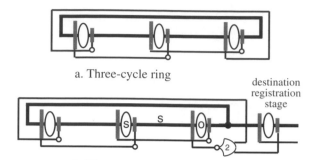

a. Three-cycle ring

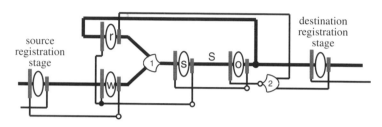

b. Three-cycle ring with output destination cycle

c. Three-cycle ring with alternate input cycles

Figure 6.1   Building the basic data path cycle structure for the ring memory element.

in the ring circulating back for storage or a new wavefront from the source registration stage. Only one wavefront at a time can flow through the fan-in node, so this fan-in structure must be conditional and must be managed with a control variable.

A control variable is added in Figure 6.2*a* with two values Write (W) and Read (R) to control the fan-in structure. For a Read, the wavefront fed back in the ring is steered into the storage cycle S. For a Write, the external wavefront is steered into the storage cycle. The behavior of the stored wavefront itself is different for each case. In Figure 6.2*b* the control values are extended to manage the behavior of the stored wavefront flowing out of the storage cycle. In the case of a Read, the stored wavefront flows through registration stage O and is steered back into the ring to the storage cycle. In the case of the Write, the stored wavefront is blocked at registration stage O and is consumed by being overwritten by the NULL wavefront to make way for the new DATA wavefront. The extension of the Write control value directly presents a request for NULL to cycle S, allowing a NULL wavefront to overwrite the stored DATA wavefront. For a Write operation cycle S becomes a consume cycle.

In Figure 6.2*c* the control variable is extended through a registration stage into its delivery pipeline, and the acknowledge paths are added to close the cycles of the control variable. Notice how the inversions (open circles) are distributed in the acknowledge paths to ensure that there is one inversion for each cycle.

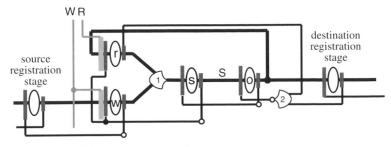

a. Control variable values are added to manage the fan-in

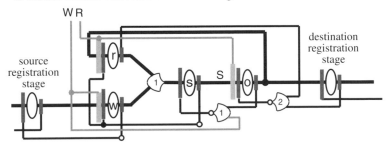

b. Control variable values are extended to manage the stored wavefront flow

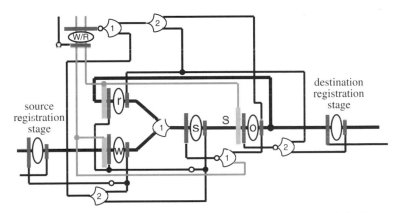

c. Acknowledge stuctures are added to close the cycles of the control pipeline

Figure 6.2   Control structure of a ring memory element.

The control variable ensures that Write and Read operations are mutually exclusive and will never conflict. There can be zero or any number of Reads following each Write. When a Write is commanded, the next wavefront arriving through the source cycle will be accepted into the ring. The Nth Write control wavefront will steer an Nth data wavefront through the source cycle to the storage cycle. The Nth Read control wavefront will deliver the stored wavefront as the Nth output wavefront from the register.

6.2 COMPLEX FUNCTION REGISTERS

Complex functional registers can be composed from this basic model.

6.2.1 A Program Counter Register

Figure 6.3 shows a next instruction address register (program counter) for a RISC architecture. The behavior of this register is slightly different from the basic

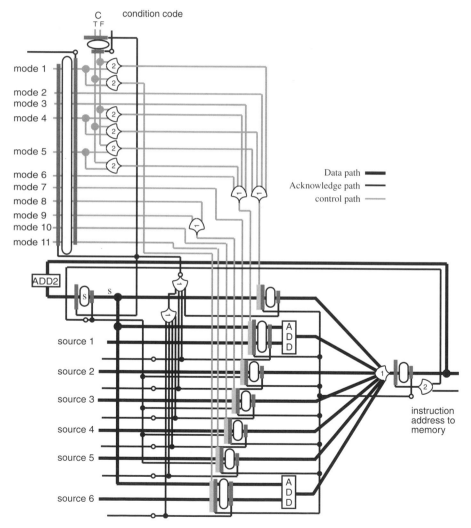

Figure 6.3 A program counter register with several branch modes and several branch address sources.

model in that there are not multiple Reads between writes. Every Read is followed by a Write. This allows a somewhat simpler control structure. There are 11 different branch modes represented by an eleven value control variable. Three of these modes are dependent on a condition code. Each decoded instruction will assert one mode variable with one value asserted. There are six sources of branch addresses such as memory, the register file and other registers, and two of these are added to the address wavefront currently stored in the ring.

This is basically a ring register with a big fan-in structure. The primary issues for this structure are completeness relationships in relation to the stored wavefront and orphan paths in relation to the combinational ADD expressions.

Completeness Issues. The fan-in structure to the ring is a varying fan-in structure. The fan-in might include the stored wavefront, an external wavefront, or both when the external wavefront is added to the stored wavefront. In this case the current address wavefront stored in cycle S is always either used or consumed to make way for a new address wavefront. So all of the selection registration stages will acknowledge the registration stage S. For the three selections that use the stored wavefront, the acknowledge is a normal wavefront flow acknowledge. For the four selections that do not use the stored wavefront, the acknowledge overwrites the stored wavefront performing a consume. A new wavefront will enter the ring and be stored in cycle S.

The selections that do not use the stored address wavefront must nevertheless ensure that the wavefront is properly stored in cycle S before asserting the acknowledge to consume it. To indicate the S storage is complete, the completion of the S registration stage is presented as a spanning enable to the selection registration stages that do not use the stored wavefront. This ensures that the current address wavefront is stored in cycle S when an acknowledge is generated to overwrite and consume it.

Orphan Issues. The ADD2 combinational expression is placed in the cycle before storage cycle S to avoid wavefront propagation issues with the circulating wavefront. If the ADD2 were in the path S, the whole combinational expression would become an orphan path for the case of a consume cycle. Placed in the previous cycle of the ring, it is an always effective component of the data path and hence is never an orphan path.

The ADD expressions in the fan-in structure are placed after the select registration stages to avoid orphan problems. If the ADDs were placed before the select registration stages, they would be on orphan branch paths, and the whole ADD expression would be an orphan path every time its registration stage is not selected. Placed after the select registration stages, they are presented only with effective wavefronts and are never orphan paths.

6.2.2   A Counter Register

The counter register illustrates a register with multiple output destinations. The counter register counts down from a preloaded number and, with each count, outputs

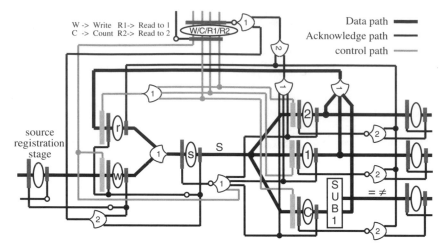

Figure 6.4 Counter register with multiple output paths.

whether the count is equal to zero or not equal to zero. The initial count value can be written from one source, and the current count value can be Read out to two destinations. The counter register of Figure 6.4 illustrates the fan-out steering structure for multiple output destinations.

The stored wavefront is steered to one of three destinations as well as consumed for the Write operation. For the count operation, the stored wavefront number is decremented and recirculated in the ring. The equal/not equal variable is presented to an output path. For Read 1 and Read 2 the stored wavefront is steered to the appropriate output path and is recirculated without change.

6.3 THE CONSUME/PRODUCE REGISTER STRUCTURE

Another approach to implementing a register, somewhat more efficient in both energy and hardware than circulating the wavefront in a ring, is to stop the wavefront and sample it as desired. This involves an auto consume cycle with an inverted phase relationship. For most cycles in a system the stable idle state is NULL, and an occasional DATA wavefront flows through it. For the phase inverted cycle the stable idle state is DATA, and an occasional NULL wavefront flows through it.

Figure 6.5 shows a stretched out graphic form of the register and a corresponding 2NCL expression for a data path of a single 4 value variable. The register consists of a phase inverted storage/consume registration stage (SC) followed by a produce registration stage (& and P), both coordinated by a 2 value control variable. While there is a data path (S) between the storage/consume and produce registration stages, it is not part of a cycle, and there is no direct coordinating interaction between the two registration stages. The two registration stages do not form a cycle. Instead, the single control variable spans the two registration stages and

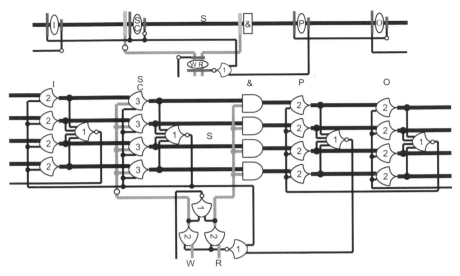

Figure 6.5   Consume/produce register structure.

coordinates their behavior. The control pipeline presents a sequence of Reads or Writes one at a time. The NULL control wavefront ensures that each operation is completed before the next operation is begun and that they never interfere.

Figure 6.6*a* shows the complete register structure and its input and output pipelines. Figure 6.6*b* shows the Write control cycle. Notice that there are three inversions in the phase-inverted Write control cycle. Figure 6.6*c* shows the Write input cycle. Figure 6.6*d* shows the auto consume cycle of registration stage SC. Figure 6.6*e* shows the Read/produce cycle.

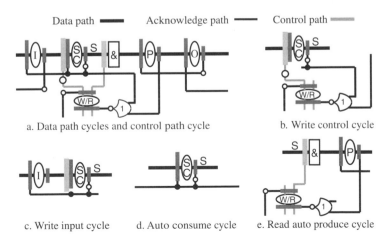

Figure 6.6   Cycle structure of the consume/produce register.

A wavefront is captured and stably maintained on path S by the SC registration stage. It is read by producing a new data wavefront and NULL wavefront from the stored wavefront via the produce registration stage.

6.3.1 The Read Cycle

The Read wavefront is produced through the Boolean AND operators (&) via the stable wavefront stored on path S and the Read control value. This is one of the very few places in logically determined design where the Boolean AND operator is useful and benign. In Figure 6.7a the register begins in a stable state with an input data wavefront and waiting for a control variable wavefront. In Figure 6.7b a control variable arrives with a Read value. This enables the DATA wavefront values through the rank of AND operators, producing a DATA wavefront that flows into the output pipeline. The data wavefront flows through registration stage P, which requests NULL from the control registration stage W/R.

In Figure 6.7c the control variable NULL wavefront arrives and forces a NULL wavefront via the rank of AND operators. A DATA-NULL wavefront sequence has been produced from the stable wavefront on the storage path S. In Figure 6.7d the NULL wavefront propagates through the output pipeline and the register becomes stable again, waiting for the next control variable wavefront.

6.3.2 The Write Cycle

The storage/consume registration stage (SC) will capture a wavefront and stably maintain the wavefront's values. Notice that the Write control value presented to the SC registration stage is inverted and that the acknowledge/request signal from the SC registration stage to the control registration stage is inverted twice. This effectively

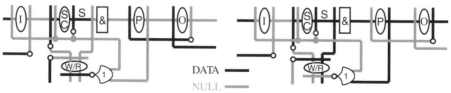

a. Stable waiting for control variable

DATA ——
NULL ——

b. Control variable arrives with READ value and DATA wavefront is produced

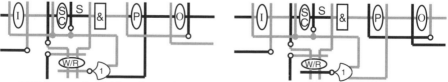

c. NULL control variable arrives and NULL wavefront is produced

d. NULL wavefront flows and the register becomes stable waiting for the next control variable

Figure 6.7   Read cycle for the auto consume/auto produce register.

inverts the phase of the registration stage in relation to the other registration stages in the system. While most other registration stages are normally maintaining a NULL wavefront the SC registration stages is normally maintaining a DATA wavefront.

The Write control variable value is normally NULL. The inverted control value presented to the SC registration stage is normally DATA and will maintain the last DATA wavefront that played through the registration stage. In Figure 6.8a the register begins in a stable state waiting for a control variable wavefront. In Figure 6.8b, a control variable arrives with a Write value. This becomes a NULL value at the input of the SC registration stage, which allows the NULL wavefront of registration stage I to flow through registration stage SC, overwriting and consuming the stored DATA wavefront.

In Figure 6.8c, registration stage SC detects the NULL wavefront and requests a DATA wavefront from registration stage I, requests a DATA wavefront for its own auto consume cycle, and requests a NULL wavefront from the control registration

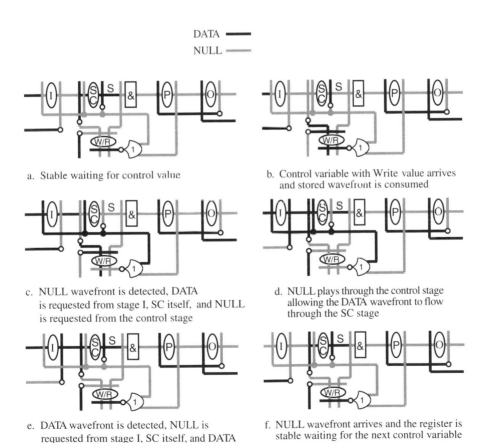

DATA ━━━
NULL ══════

a. Stable waiting for control value

b. Control variable with Write value arrives and stored wavefront is consumed

c. NULL wavefront is detected, DATA is requested from stage I, SC itself, and NULL is requested from the control stage

d. NULL plays through the control stage allowing the DATA wavefront to flow through the SC stage

e. DATA wavefront is detected, NULL is requested from stage I, SC itself, and DATA is requested from the control stage

f. NULL wavefront arrives and the register is stable waiting for the next control variable

Figure 6.8   Write cycle for the auto consume/produce register.

stage W/R. A DATA wavefront flows through registration stage I. In Figure 6.8*d*, the NULL control wavefront arrives. This is inverted to a data value at the SC registration stage that allows the DATA wavefront to flow through registration stage SC into S. In Figure 6.8*e*, registration stage SC detects the DATA wavefront and requests a NULL wavefront from registration stage I and a DATA wavefront from the control registration stage W/R. In Figure 6.8*f*, a NULL wavefront flows through I and the register enters a stable state waiting for the next control variable. Registration stage SC will maintain the DATA wavefront until another Write control value arrives.

6.4 THE REGISTER FILE

A register file can be constructed by placing individual registers in a fan-out–fan-in structure. The basic register file structure is shown in Figure 6.9. Three register elements are integrated into a fan-out–fan-in steering structure. A wavefront can be written to the selected register by directing it from the input data path through the appropriate fan-out path and commanding a Write operation. A register can be read by commanding a Read operation directing the output from the appropriate fan-in path to the output data path.

The SC registration stage of each register is exposed to the fan-out bus that is continually transitioning. If the only input to registration stage SC were the control value and the data path and all the control values are DATA, then each data wavefront on the bus would flow through all the SC registration stages and overwrite all the stored wavefronts. What protects each SC registration stage and its stored

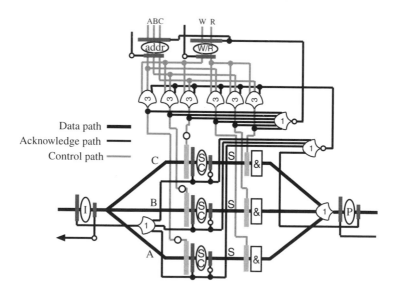

Figure 6.9 Sequential access register file.

wavefront is the NULL value of the auto-consume input to the SC registration stage. To isolate a registration stage on a transitioning bus, there must be two control inputs asserting different values. If there is one control input asserting NULL, it will allow NULL wavefronts through. If there is one control value asserting DATA, it will allow DATA wavefronts through. If there are two control inputs asserting different values, then the registration stage becomes locked in a noncompeteness state, so no data path wavefronts can flow through the registration stage. Notice that in Figure 6.8a the two control inputs to the SC registration stage in the stable waiting state are different values. This is the reason for the auto-consume cycle around the SC registration stage.

In this instance there is one 3 value address variable and one 2 value control variable. The address variable and the 2 value control variable are combined into a single 6 value variable that controls the register file. The behavior of the register file is sequentially ordered by the sequence of control variable wavefronts.

6.4.1   A Concurrent Access Register File

The key to concurrent behavior is distributed control. There is no particular need to combine all the control of the register file into a single control variable. The input address, the output address and the control of each register can be independently generated and resolved. There is an Nth fan-out address associated with each Nth Write control variable value and an Nth fan-in address associated with each Nth Read control variable value. Figure 6.10 shows a register file with distributed control. Each register is embedded in a pipeline that can store one input wavefront and one output wavefront. The fan-out control can direct a wavefront to a register. The wavefront flows into the pipeline and is presented to the I registration stage awaiting the Write command. In the meantime the fan-out control can direct the next wavefront over a different fan-out path to a different register. If the next wavefront is directed to the same register, the fan-out structure blocks and waits until the Write is completed for the previous wavefront; at this time the blocked wavefront will flow into the register pipeline and free the fan-out structure for another operation. If the input data path and fan-out structure are much faster than the register elements, then many register elements can be operating concurrently while not slowing down the input data path.

Since the SC registration stage is not exposed to the ambiguity of the fan-out path, the auto-consume cycle around SC is not needed here. Registration stage I blocks each next wavefront from the SC registration stage. The isolation on the fan-out bus is expressed with the select control values that are always different from the acknowledge value for the registration stage until the registration stage is selected.

The Read operations also buffer their output in the register element pipeline awaiting the appropriate address to steer them out of the register file structure. So, if the output data path and fan-in structure is much faster than the Read operation, then multiple registers can be concurrently performing a Read operation. The slower Read operations will not slow down the faster output data path.

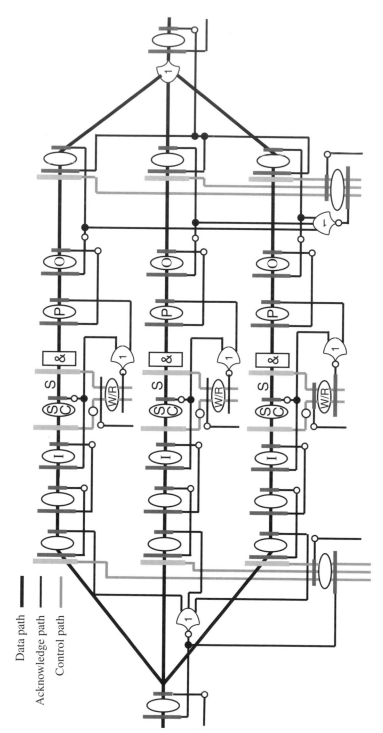

Data path ▬▬▬
Acknowledge path ───
Control path ───

Figure 6.10 Register file with distributed, concurrent control.

Again, if there are consecutive Reads from the same register, the whole structure waits on the wavefronts from the Reads.

The Write and Read operations are strictly sequenced for each individual register but are quite independent among different registers, so one register can be read while another register is being written.

6.4.2   2D Pipelined Register File

The register file can be 2D pipelined just like any other data path function by pipelining the control variables across the data path (see the discussion on two-dimensional pipelining in Section 5.3 of Chapter 5). The pipelining can be MSV first or LSV first. Write wavefronts will flow in diagonally, Read wavefronts will flow out diagonally, and if a Read immediately follows a Write, a wavefront can flow diagonally right through the register file.

6.5   DELAY PIPELINE MEMORY

A wavefront may be delayed N wavefronts in relation to itself through a parallel pipeline in which N wavefronts were initialized. It will then continually store N wavefronts. Figure 6.11 shows such a delay structure. The wavefront from registration stage I goes over the upper data path to registration stage D1 and over the lower data path into the B input of the combinational expression. The A input, presented from the upper data path, is a delayed wavefront. For every wavefront that flows through registration stage I and over the lower pipeline through the combinational expression, one wavefront will enter the upper pipeline, and one wavefront will leave the upper pipeline. The wavefronts in the upper pipeline will always be delayed by the number of wavefronts that were initialized in the upper pipeline.

6.6   DELAY TOWER

A pipeline delay can be expensive in terms of switching cost. Every wavefront flows through every cycle in the pipeline. For large delays the delay tower of Figure 6.12 with a fan-out–fan-in structure can be much more efficient. Each data path will store

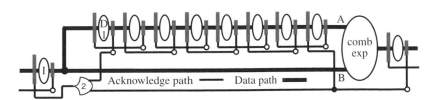

Figure 6.11   Pipeline delay line.

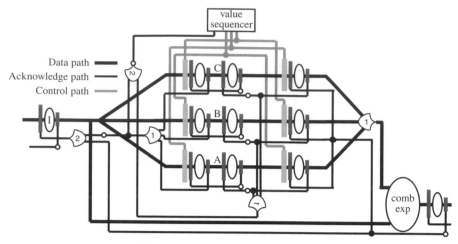

Figure 6.12   Delay tower structure.

one wavefront and each wavefront only propagates through one data path. When the data path is selected for a Write, the wavefront flows into a pipeline and then waits to be selected for a Read to flow out of the pipeline.

The value of the control variable rotates Writes and Reads through the data paths. The control values skew the Writes and Reads such that each Write is one path behind the Read in the rotation of paths. The result is that the structure is storing two wavefronts and the Read output wavefront is always two ahead of the Write input wavefront. The rotation sequence is Read A – Write B, Read C – Write A, Read B – Write C. The structure must be initialized with a DATA wavefront in pipeline A and a DATA wavefront in pipeline C.

6.7   FIFO TOWER

While a logically determined pipeline is inherently a first-in–first-out memory structure (FIFO). It can, again, be more efficient in terms of switching to use the tower structure. In the FIFO tower of Figure 6.13 the Read and Write operations are controlled by independent sequencers. The Reads and Writes free run in relation to each other. When the FIFO is empty, the next Read will wait on a DATA wavefront to arrive. When the FIFO is full, the next Write will wait on a bubble to arrive. The structure does not need to be initialized with wavefronts. The two sequencers should begin, enabling the same pipeline to Write and Read.

6.8   STACK TOWER

See the discussion on stack controller in Section 7.1.4 of Chapter 7.

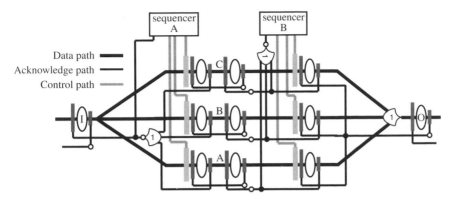

Figure 6.13   Tower FIFO structure.

6.9   WRAPPER FOR STANDARD MEMORY MODULES

Figure 6.14 shows a wrapper that allows the use of a standard binary memory in a logically determined system. The wrapper presents a boundary of logically determined interaction protocol while accommodating the timing issues related to the data format conversions and the behavior of the memory itself inside the boundary.

6.9.1   The Write Operation

A Write operation specified by a Write value of the control variable waits until there is a complete data wavefront and a complete address wavefront. A Write command is then presented to the memory and a delay is exercised. After the delay all three input registration stages are acknowledged indicating the Write operation is complete. The NULL wavefronts remove the Write command and acknowledge

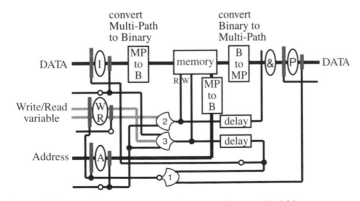

Figure 6.14   Logically determined wrapper for standard binary memory.

the input registration stages with a request for a DATA wavefront, resetting the wrapper for another operation.

6.9.2 The Read Operation

A Read operation specified by a Read value of the control variable waits until there is a complete address wavefront. A Read command is then presented to the memory, and a delay is exercised. After the delay the produce registration stage is enabled, and a DATA wavefront flows through the output register P. The acknowledge from P and the completion of the delay acknowledge the input registration stages with a request for NULL. The NULL wavefront then removes the Read command and produces a NULL wavefronts through the produce stage that flows through the P registration stage. The input registration stages are acknowledged with a request for a DATA wavefront resetting the wrapper for another operation.

6.9.3 The Binary Conversions

The binary conversions, because they involve binary signals, cannot be logically determined. They are Boolean functions, and their timing behavior must be considered in the delays associated with the memories. The conversion expressions for both dual-path and four-path encodings are shown in Figure 6.15

6.9.4 2D Pipelined Memories

The wrapper and the memory can be 2D pipelined. The wrapper can be 2D pipelined in either direction. The memory must be partitioned into independent slices, each with full address decode for each slice, matching the granularity of the data path partitioning. This may or may not be feasible. The alternative is to provide triangular buffers on the input and output data paths of the memory to convert between diagonal wavefronts and vertical wavefronts.

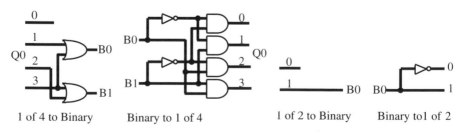

1 of 4 to Binary   Binary to 1 of 4   1 of 2 to Binary   Binary to1 of 2

Figure 6.15   Binary conversion expressions.

6.10 **EXERCISES**

6.1. Design and simulate a 2D pipelined register file.

6.2. Design and simulate a wrapper and memory for a 2D pipelined memory.

6.3. When does it become more efficient to use a delay tower instead of a pipeline delay?

6.4. Design and simulate a one-dimensional 5 sample FIR filter and IIR filter.

CHAPTER 7

State Machines

In a logically determined system the state is expressed as spontaneously flowing wavefronts. Wavefront flow can be constrained in a ring to express the localized behavior of a state machine.

7.1 BASIC STATE MACHINE STRUCTURE

The basic structure of a logically determined state machine, illustrated in Figure 7.1, is a ring coupled to a pipeline through a combinational expression. The ring maintains the current state. The pipeline presents input wavefronts and accepts output wavefronts. The combinational expression receives the current state and the next data input and asserts the next state and a data output.

The ring must be initialized with one DATA wavefront expressing the initial state. If the ring is initialized to all NULL, there will be no current state wavefront for the first DATA wavefront, and the structure will deadlock. The ring will always contain exactly one DATA wavefront and one NULL wavefront.

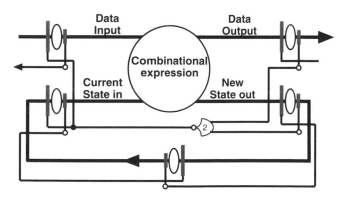

Figure 7.1   Basic state machine structure.

Logically Determined Design: Clockless System Design with NULL Convention Logic^{TM}, by Karl M. Fant
ISBN 0-471-68478-3   Copyright © 2005 John Wiley & Sons, Inc.

The combinational expression has a fan-in of two inputs and a fan-out of two, so the acknowledge signals completing its cycle have a fan-in of two and a fan-out of two. The completeness criterion of the combinational expression ensures that the Nth state wavefront in the ring will interact with the Nth DATA wavefront propagating along the data path. The output of the combinational expression is the data wavefront presented to the pipeline and the new state wavefront presented to the ring.

7.1.1 State Sequencer

The simplest state machine is a state sequencer, and it does not have a combinational expression. There are no input data, just output data and the next state. The state values are just shifted within the state variable to transition to the next state. Figure 7.2 shows a 4 value state sequencer implemented as a 4 cycle ring with a 4 value variable. The ring presents its output to a pipeline and recirculates the next state in the ring. A next state variable wavefront is delivered with each acknowledge requesting data.

7.1.2 Monkey Get Banana

The monkey get banana machine is a state machine that controls a banana vending machine for a monkey. The monkey has to push 4 buttons—A, B, C, D—in sequence to get a banana. If she pushes any button out of sequence, the machine is reset and she has to start with A again, Figure 7.3 shows the state diagram and the state function map. The buttons are expressed as a 4 value variable, and the states of the machine are expressed as a 4 value variable. The combinational expression can be read directly off the function map. The combinational expression in this instance receives signals from a cycle that intersects the outsides world. For the state machine to work, the button signals must behave as proper wavefronts. For instance, two button cannot send signals simultaneously, and there must be a NULL wavefront between each

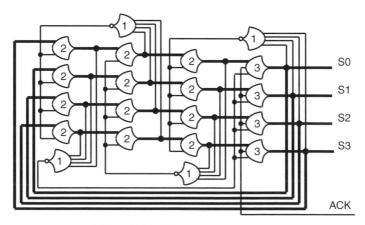

Figure 7.2 Simple state sequencer.

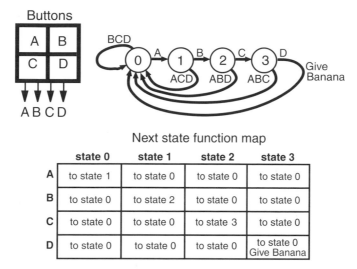

Next state function map

	state 0	state 1	state 2	state 3
A	to state 1	to state 0	to state 0	to state 0
B	to state 0	to state 2	to state 0	to state 0
C	to state 0	to state 0	to state 3	to state 0
D	to state 0	to state 0	to state 0	to state 0 Give Banana

Figure 7.3   Monkey get banana state control.

button signal. These conditions can be accomplished by conditioning the signals with an arbiter. See the discussion on 'the arbiter' in Section 8.4.2 of Chapter 8.

The output is the delivery of a banana. There is no acknowledge from the monkey. The machine just dumps a banana and goes to state 0.

7.1.3   Code Detector

The code detector is a state machine to detect the sequence 0010111 in a sequential stream of binary variables. The combinational expression for the state machine was

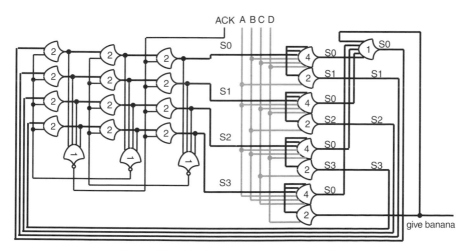

Figure 7.4   Monkey get banana state machine.

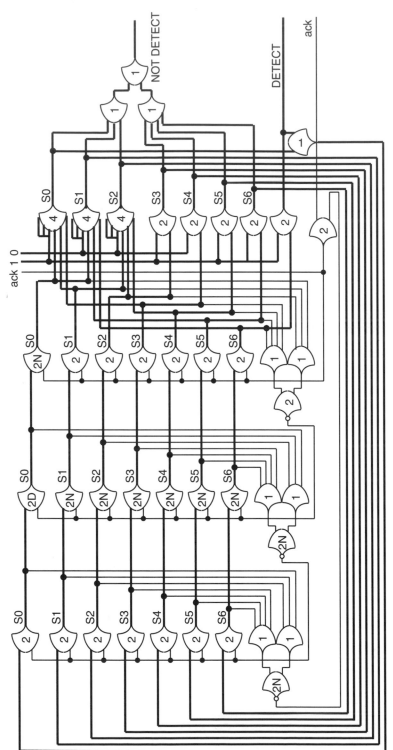

Figure 7.5 Code detector state machine.

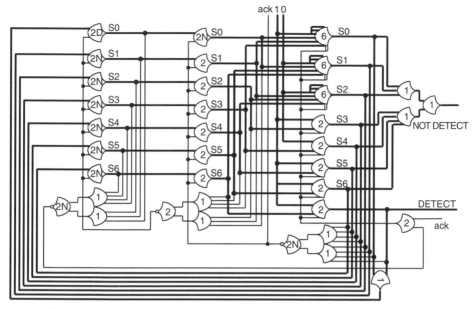

Figure 7.6 Detector state machine with embedded combinational expression.

presented in Section 4.8 of Chapter 4. Here the combinational expression is inserted in a 3 cycle ring in Figure 7.5 to form the state machine. The ring maintains the 7 value state variable. Initializing operators are shown to initialize a state wavefront to state 0 in the ring. The initializing signal is not shown. The state variable is

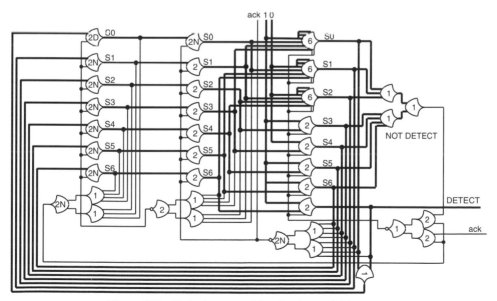

Figure 7.7 Code detector with a 1 value variable output.

combined with the binary input variable through the combinational expression to determine the next state and a 2 value decision variable with the meanings 'detect' and 'not detect'.

The state machine can be optimized by making the combinational expression a registration stage of the ring as shown in Figure 7.6. Adding the acknowledge/ request to the combinational rank in this instance results in three 5 input operators that may not be available in a library. This is an instance that arises frequently, and it might be advantageous to define special 5 input operators that can receive the one extra necessary signal, the acknowledge, to allow the optimization of combinational expressions into registration stages.

Single-value Detect Variable. With a 2 value detect variable the destination of the detect variable has to continually acknowledge the state machine for both 'detects' and 'not detects'. It might be convenient if the detect wavefront is generated only when there is a 'detect'. Consequently there is an acknowledge only when a detect wavefront is generated. The state machine must be able to determine when to expect an acknowledge and when to not expect an acknowledge. This is determined by the 8 value variable generated by the combinational expression, so the acknowledge completeness must be conditioned by this variable.

The strategy in this case is to fill in the acknowledges so that the state machine always receives an acknowledge. The seven values of the output variable that do

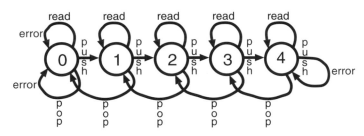

a. State machine diagram

	state 0	state 1	state 2	state 3	state 4
read	ERROR to state 0	READ to state 1	READ to state 2	READ to state 3	READ to state 4
pop	ERROR to state 0	READ to state 0	READ to state 1	READ to state 2	READ to state 3
push	WRITE to state 1	WRITE to state 2	WRITE to state 3	WRITE to state 4	ERROR to state 4

b. Next state function map

Figure 7.8   Stack controller state diagram and function map.

not generate a detect wavefront are collected to generate what is essentially a 'not detect' acknowledge. Essentially the 'not detect' value is auto-consumed. When there is a 'detect', there will be an acknowledge from the receiving cycle of the detect wavefront. The resulting expression is shown in Figure 7.7.

7.1.4 Stack Controller

The stack controller manages a 4 entry stack with 'push', 'pop', and 'read', commands. The read command reads the top of the stack without popping it. The controller implements the stack by sending addresses and commands to a register file (see the discussion of 'the register file' in Section 6.4).

The state machine maintains a 5 value state variable that represents the current top of the stack. States 1 through 4 are direct address lines for the register file.

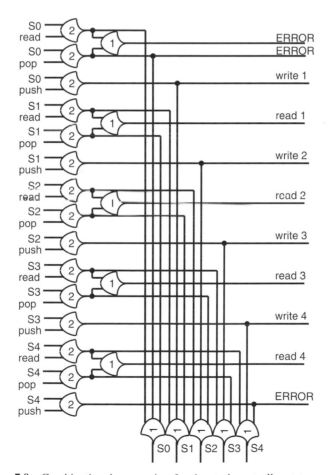

Figure 7.9 Combinational expression for the stack controller state machine.

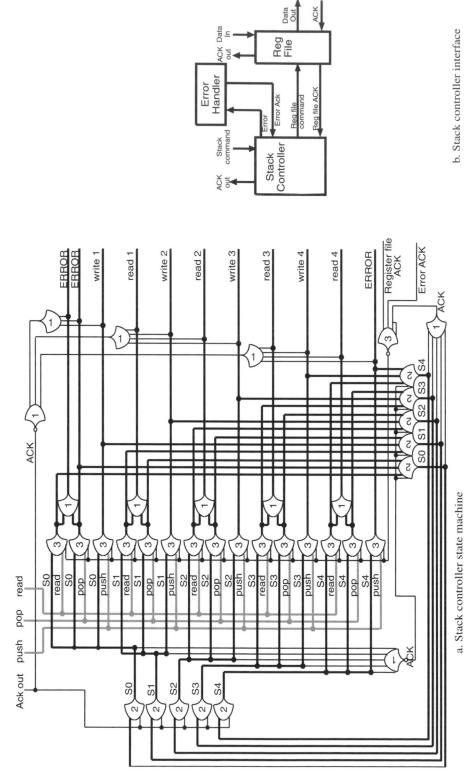

Figure 7.10 Stack controller state machine and interface diagram.

a. Stack controller state machine

b. Stack controller interface

State 0 is the empty state. The input to the state machine is a three value command variable specifying 'push', 'pop', or 'read'. For each command it sends an address and command to the register file and sets the next state. For the empty and full conditions it outputs an error code. The input to the stack controller is a 5 value state variable and a 3 value command variable. The state diagram is shown in Figure 7.8a, and the state function map is shown in Figure 7.8b.

The combinational expression for the controller can be read directly from the function map and is shown in Figure 7.9. The first rank of logic generates the commands to the register file, and the second rank generates the next state. Each rank of logic can be a registration stage. The complete state machine is shown in Figure 7.10a. It is a ring of three cycles. Two of the registration stages are formed from the two ranks of logic, and only one is serving just as a registration stage.

The complete stack is shown in Figure 7.10b. The combinational expression always generates an internal wavefront to the next ring registration stage and always generates an external wavefront to the register file or to the error handler, so there must be an acknowledge from the next ring registration stage and an acknowledgement from either the register file or the error handler.

7.2 EXERCISES

7.1. What are the problems of a nondeterministic interface with the real world. Discuss the relationships between the monkey and the get banana state machine. Can the behavior of the monkey be integrated into the cycle structure of the machine?

7.2. Design an 6 bit code detector that will detect any 6 bit code.

7.3. Design and simulate a 2D pipelined stack.

CHAPTER 8

Busses and Networks

Data path wavefronts are steered through a logically determined system by interacting with control wavefronts. The basic fan-out and fan-in steering structures have already been presented in Section 3.3 of Chapter 3. This chapter will elaborate on larger structures built from those basics.

8.1 THE BUS

8.1.1 The Serial Bus Structure

A bus steers wavefronts from multiple source paths to multiple destination paths. The most familiar form is the serial bus over which one wavefront can flow at a time. Figure 8.1 shows a serial bus with two source paths and five destination paths. The two source paths fan-in to a single path, which then fans-out to the five destination paths. The source registration stages are controlled by a single 2 value control variable. The destination registration stages are controlled by a single 5 value control variable. The Nth data wavefront will be steered by the Nth source control variable and the Nth destination control variable.

The throughput of a serial bus can be increased with 2D pipelining. Several wavefronts can simultaneously flow through the bus as the control variables propagate across the data path. As with most 2D pipelined structures, the throughput of the bus will be independent of the width of the data path. A 2D pipelined serial bus might actually be the most efficient high-throughput structure for very wide data paths.

8.1.2 The Crossbar

By fanning-out the source paths to many internal paths and fanning-in these internal paths to the destination paths, the many internal paths can support concurrency of wavefront flow through the bus. Figure 8.2 shows the crossbar structure.

Each source path and destination path is now controlled by its own control variable. It is now possible to enable path A to path 4 and to simultaneously enable path

Logically Determined Design: Clockless System Design with NULL Convention Logic[TM], by Karl M. Fant
ISBN 0-471-68478-3   Copyright © 2005 John Wiley & Sons, Inc.

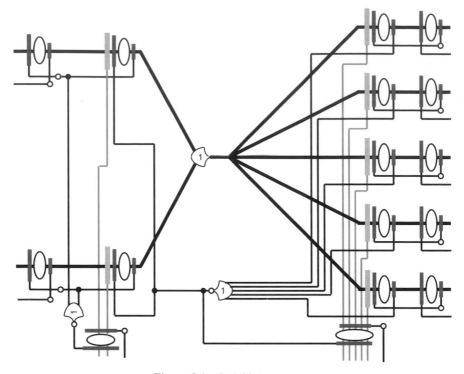

Figure 8.1   Serial bus structure.

B to path 3. If paths A and B are simultaneously enabled to path 4, the control variable of path 4 will determine the sequence of wavefronts through path 4. All possible
concurrency is automatically exploited by the structure, and all conflicts are automatically resolved by the structure itself. Again, at each stage the Nth control variable wavefront steers the Nth data path wavefront and everybody waits for
completeness.

Consolidated Fan-out Controls.   The reader may notice that there is an enable
registration stage at both ends of each internal path. There only needs to be one
enable per path as long as there is only one path enabled into each fan-in. Both control values that enable the ends of the path can be grouped at the fan-out registration
stage of each path. When the enable values are grouped, a path is enabled only if
both control values are DATA. The values A to 1 and 1 from A will enable one
path and the values B to 2 and 2 from B will enable another path. The A to 1 and
B to 1 will never be simultaneously enabled because the values 1 from A and 1
from B are mutually exclusive. The values 2 from A and 3 from A will never be simultaneously enabled because the values A to 2 and A to 3 are mutually exclusive.
Again, the structure inherently exploits all possible concurrency and inherently manages all flow conflicts. Figure 8.3 shows the crossbar with the consolidated control

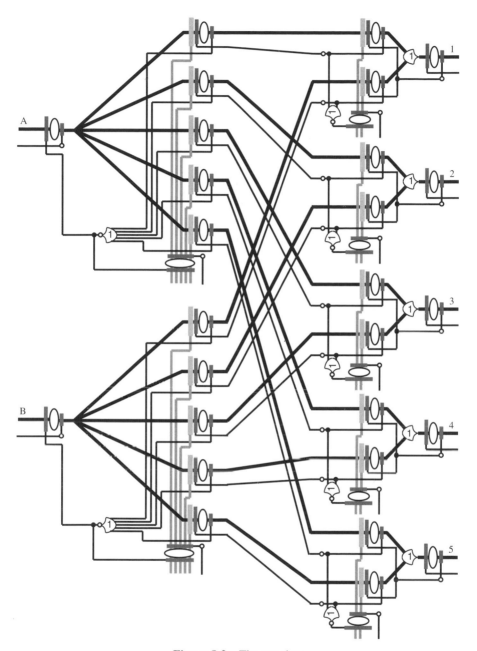

Figure 8.2 The crossbar.

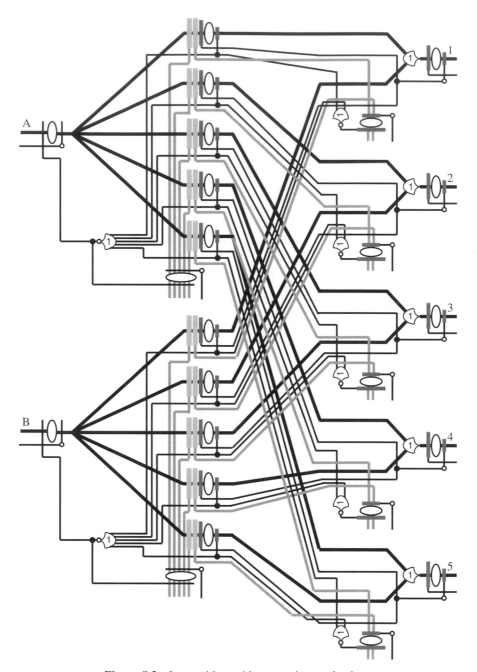

Figure 8.3 Inverted bus with grouped control values.

values. Half of the control registration stages have been eliminated by consolidating the control values.

Interweaving Control Variables. Now that the control values have been consolidated, the two values that enable each path can be combined into a single control value for each path. If one wishes to keep the distributed control of concurrent behavior that the independent control variables provided, the fact that a single combined control value belongs to two control variables must be maintained.

These shared values can be woven into a single path. Figure 8.4 shows the seven independent control variable paths combined into single interwoven control value path. The acknowledge structure for the combined path is exactly as it would be for the seven independent paths. The registration stage that enabled data path A4 would acknowledge the control registration stage for source path A and the control registration stage for destination path 4. In the interwoven path the value A4 will

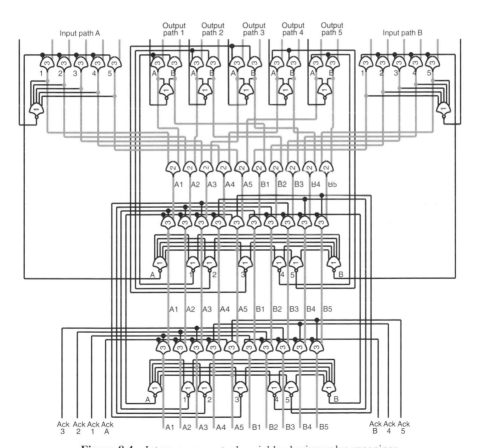

Figure 8.4　Interwoven control variable sharing value meanings.

acknowledge all values sharing A and all values sharing 4. Each single shared value is still a member of two cycles, and both cycles must be properly closed with acknowledges. If there is no conflict, control values will flow freely by each other and concurrently enter the bus structure. If there is a conflict, one value will be blocked and the conflicting values will enter the bus structure in sequence.

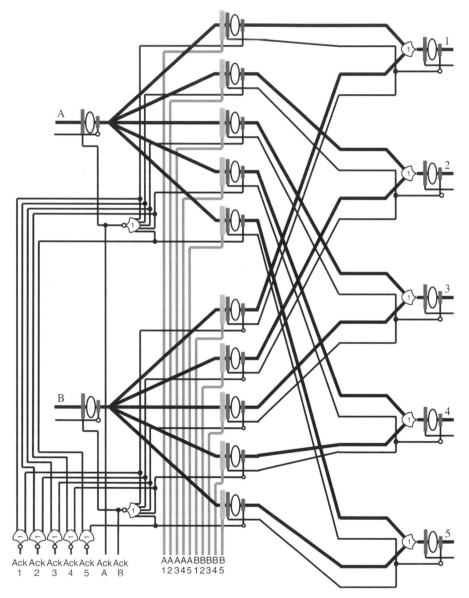

Figure 8.5 Bus with interwoven control values.

The seven control paths with 20 values has been reduced to one control path with 10 values. The bus structure using the interwoven control values is shown in Figure 8.5. The structure still exploits all possible concurrency and inherently manages all conflicts.

8.2 A FAN-OUT STEERING TREE

Wavefronts can be steered by stages through a fan-out tree. A multi-variable address propagates with each wavefront. At each stage an address variable is consumed to steer the wavefront. At the end of the steering tree the address variables have been consumed and the wavefront is delivered to the correct output data path. Figure 8.6 shows a two-stage fan-out steering tree. Two address variables steer a wavefront through the network. The first stage fans out to three paths controlled by a 3 value variable. The second 2 value address variable is steered along with the data wavefront through the first stage and controls the second-stage fan-out to two paths.

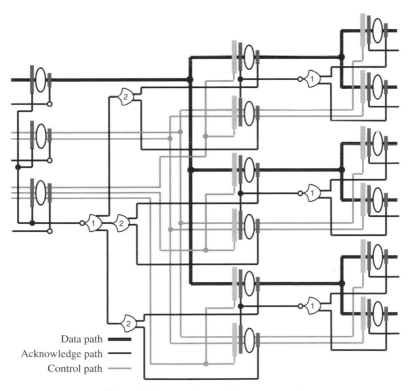

Data path ■■■■
Acknowledge path ———
Control path ———

Figure 8.6 Multi-stage fan-out network.

As wavefronts propagate concurrent propagation increases. Many wavefronts can be propagating through the tree simultaneously. A steering tree can feed wavefronts from a fast input path into concurrently propagating slower paths, which might be functional paths.

8.3   FAN-IN STEERING TREES DO NOT WORK

Fan-in structures cannot be structured as trees and must be flat. Figure 8.7 shows a fan-in tree structure. The fundamental question is which input wavefront should be steered to the output path. In the tree structure the final decision is made between two wavefronts presented to the final stage. If there are two DATA wavefronts on the internal paths just prior to the last stage and only one is selected to propagate to the output, then one DATA wavefront is marooned on an internal path. This marooned wavefront will deadlock the structure because the wavefront will never flow and the input will never receive another request for DATA.

The marooned wavefront must be consumed. If it is a valid wavefront, it is erroneous to consume it. If it is not a valid wavefront, then it must be a dummy wavefront generated specially to be consumed in order to satisfy the control structure. To generate the dummy wavefront, it must have been determinable in advance which

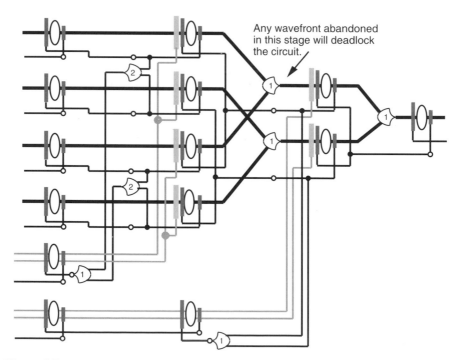

Figure 8.7   Fan-in tree network structure that demands two wavefronts on the internal paths.

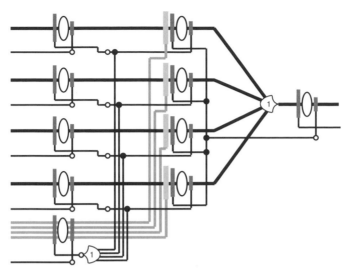

Figure 8.8   Flat fan-in structure.

wavefront is going to get to the output, so there is no point in delaying the decision. All of this adds up to the realization that fan-in steering structures cannot and should not be structured as a tree. They must be a flat, single stage structure as shown in Figure 8.8.

8.4   ARBITRATED STEERING STRUCTURES

Arbitration is required when there are simultaneous requests for a resource that can only be used by one requester at a time. In a logically determined system the wavefront flow through most resources is determined by the sequencing of data path wavefront and control path wavefronts. Conflict can occur with resources that are logically indifferent to access order. One such circumstances is when two wavefronts need to flow through the same data path segment in a switching network. If two wavefronts can try to simultaneously use a single data path, their access must be arbitrated and sequenced. One must be blocked while the other is allowed to pass. When the other has finished its passage, the one can be allowed to proceed.

8.4.1   The MUTEX

Arbitration begins with a mutual exclusion operator or MUTEX. A MUTEX operator of Figure 8.9 receives two request signals as inputs and outputs strictly one at a time of two grant signals. If R1 and R2 simultaneously transition high at the input of the, The MUTEX will allow only one signal through as G1 or G2 transitioning high. The two input signals contest for passage and only one wins. If one signal is ahead of

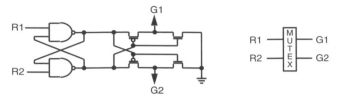

Figure 8.9   CMOS MUTEX.

the other, then it wins and the later signal is blocked. Assume that R1 won and that G1 high is asserted by the MUTEX. The MUTEX will continue asserting G1 high and blocking R2 until R1 transitions low, removing the request. Upon R1 transitioning low, the MUTEX will immediately transition G1 low and allow R2, if it is present, by transitioning G2 high. A MUTEX does not alternate inputs. If there are 5 consecutive R1 requests, there will be 5 consecutive G1 grants.

If the two requests transition simultaneously, the resolution of the conflict can take an arbitrarily long time. This condition of indecision is called metastability. While the duration of metastability is a critical issue for clocked systems, it is less so for logically determined systems. While it may create a throughput issue, it does not create a correctness of logical operation issue. The logical will wait as long as necessary for the metastability to resolve.

The MUTEX and metastability are well-studied topics [2,13]. In-depth discussions of mutual exclusion and arbitration can be found in recent texts [33,52,59].

8.4.2   The Arbiter

In arbitrating control behaviors the immediate grant of the other request by the MUTEX, when one request is removed, may not adequately serve the desired behavior protocol. One might wish, for instance, to have more positive control over when the next grant occurs. An arbiter is an expression wrapped around a MUTEX to provide this extra control. A standard form of arbiter known as a request/grant arbiter is shown in Figure 8.10.

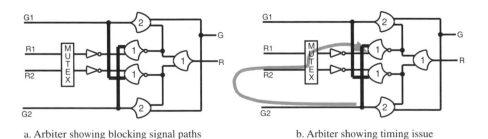

a. Arbiter showing blocking signal paths   b. Arbiter showing timing issue

Figure 8.10   Request/grant arbiter expression.

The arbiter features crossblocking signals, the wide paths in Figure 8.10a, to allow each grant to further delay the other grant. When G1 occurs the crossblocking signal blocks the grant of a G2 request. G1 will eventually lead to the release of the R1 request, whereupon the MUTEX can immediately grant G2. The G2 grant is blocked, however, until request R is released, grant G is released, and the blocking signal is released. Thus the granting of the next request can be controlled by the external G signal.

There is a timing issue shown in Figure 8.10b. The blocking signal must effectively propagate before the MUTEX is able to transition its grant. When G2 is asserted, the blocking signal must propagate and block before G2 causes R2 to be released and allows R1 to be granted. From the branch of G2, the blocking signal must be strictly faster than the path indicated by the gray highlight of Figure 8.10b. The added control around the MUTEX allows NULL wavefronts to be inserted between MUTEX grants.

8.4.3   An Arbitrated 2 to 1 Fan-in

To show how the arbiter fits into a cycle structure, it is redrawn in Figure 8.11a and juxtaposed with an arbitrated 2NCL fan-in structure in Figure 8.11b. The G paths

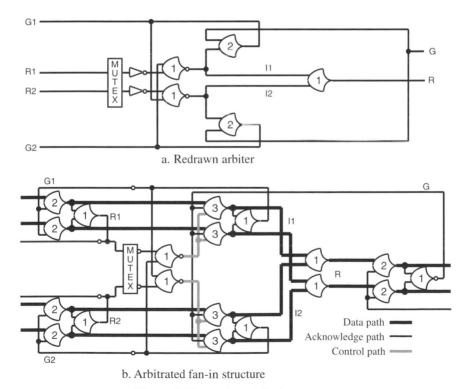

a. Redrawn arbiter

b. Arbitrated fan-in structure

Figure 8.11   Arbitrated fan-in structure.

in the arbiter correspond to acknowledge paths in the cycle structure. The R paths of the arbiter correspond to data paths of the cycle structure. The I1 and I2 paths of the arbiter correspond to the data paths through the enabled registration stages.

The presence of a complete DATA wavefront on an input stage generates a request to the MUTEX. When the MUTEX grants the request, it generates a control variable to allow the wavefront to proceed.

The MUTEX expression can be viewed as a combinational expression in the cycle except for the crossblocking paths, which essentially short-circuit the cycles. These paths have timing issues that are more critical than an orphan path. Because of this the behavior of the structure is discussed is some detail. The passage of data wavefronts through both input paths are followed through a step by step simulation.

8.4.4   Arbiter Simulation

In the simulation sequence black is DATA and gray is NULL. In step 1 of Figure 8.12, the structure is in an all NULL state awaiting DATA wavefronts. In step 2 of Figure 8.13, DATA wavefronts arrive on path A and path B simultaneously, and their completion signals generate simultaneous requests to the MUTEX. The MUTEX chooses path A and generates the grant that becomes the enabling control value for the path A control registration stage. In step 3 in Figure 8.14, the wavefront in path A flows through the path A control registration stage. It is detected and NULL is requested from the A input stage. The crossblocking signal is set to block a grant for path B.

In step 4 of Figure 8.15, the NULL wavefront propagates through the A input registration stage, is detected, and the request for path A is removed from the MUTEX. The MUTEX immediately grants the path B request, but it is blocked by the crossblocking signal. The DATA wavefront propagates through the output registration stage, is detected, and NULL is requested from the fan-in control registration stages.

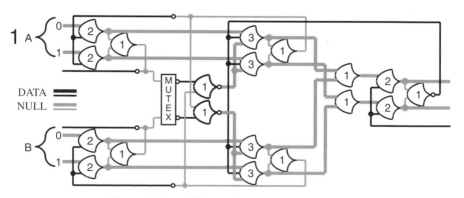

Figure 8.12   Arbitrated fan-in simulation step 1.

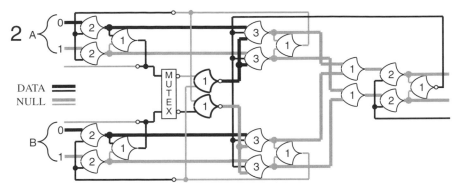

Figure 8.13 Arbitrated fan-in simulation step 2.

In step 5 of Figure 8.16, the NULL wavefront propagates through the path A control registration stage and requests DATA from the A input registration stage. At the same time the blocking signal is removed, and the grant for B is allowed, becoming the enable control value for the path B control registration stage. But the output registration stage is still requesting NULL from the control registration stages, so the DATA wavefront in path B is still blocked.

In step 6 of Figure 8.17, the NULL wavefront plays through the output registration stage and the output registration stage requests DATA from the fan-in control registration stages. A DATA wavefront plays through the A input registration stage and generates a request to the MUTEX, which is blocked by the MUTEX because it is still granting the path B request. Finally the path B DATA wavefront is enabled to flow through the path B control registration stage.

In step 7 of Figure 8.18, the B path DATA wavefront flows through the path B control registration stage, and NULL is requested from the B input registration stage. The crossblocking signal is also set to block a grant for path A.

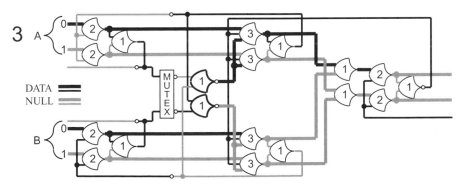

Figure 8.14 Arbitrated fan-in simulation step 3.

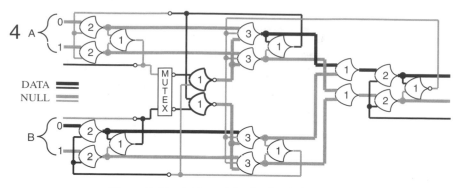

Figure 8.15   Arbitrated fan-in simulation step 4.

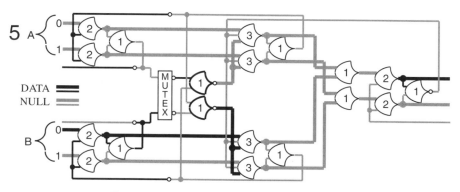

Figure 8.16   Arbitrated fan-in simulation step 5.

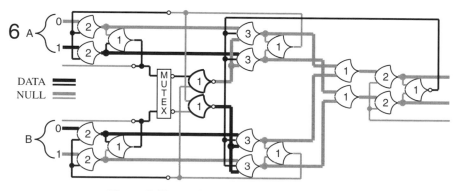

Figure 8.17   Arbitrated fan-in simulation step 6.

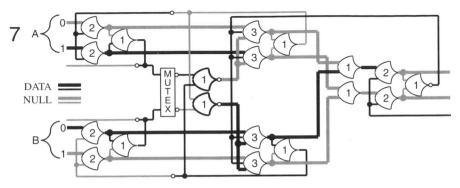

Figure 8.18   Arbitrated fan-in simulation step 7.

In step 8 of Figure 8.19, the NULL wavefront plays through the B input registration stage, removing the path B request from the MUTEX. This immediately grants the path A request, which is blocked by the crossblocking signal from path B. The DATA wavefront plays through the output registration stage and requests NULL from the control registration stages.

In step 9 in Figure 8.20, the NULL wavefront plays through the path B control registration stage. It requests DATA from the B input registration stage and releases the crossblocking signal allowing the path A control value. The DATA wavefront is still blocked, however, because the output registration stage is still requesting NULL from the control registration stages.

In step 10 in Figure 8.21, the NULL wavefront plays through the output registration stage, and DATA is requested from the control registration stages allowing the path A DATA wavefront to flow. A data wavefront propagates through the B input registration stage, and a request is presented to the MUTEX, which blocks it because it is still granting the A request. And so it goes.

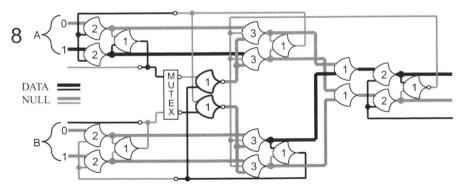

Figure 8.19   Arbitrated fan-in simulation step 8.

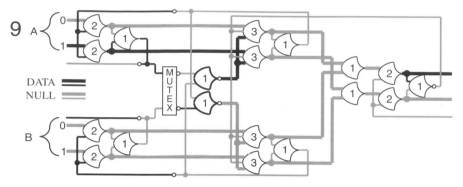

Figure 8.20 Arbitrated fan-in simulation step 9.

If there is no conflict, wavefronts will play straight through the arbiter with no waiting.

8.4.5 Arbiter Timing Issues

When the blocking path transitions to block, it must propagate from the branch marked A in Figure 8.22*a* to the threshold 1 gate faster than the path indicated by the gray highlight. Otherwise, the next grant may not be blocked.

When the blocking path transitions to unblock, there is an issue with the fan-in control stages. It must be ensured that the previous wavefront propagation is complete and that there is a NULL bubble in the control registration stages before another data wavefront can be accepted. The request from the output registration stage must cycle through NULL before another data wavefront can be enabled through the control registration stages. So it must be ensured that the NULL request from the output registration stage is present before the blocking signal allows the control value to be generated. In Figure 8.22*b* the acknowledge path must propagate

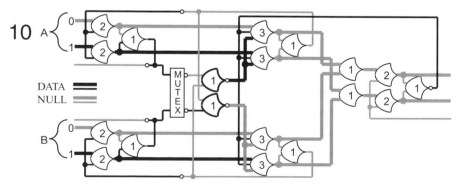

Figure 8.21 Arbitrated fan-in simulation step 10.

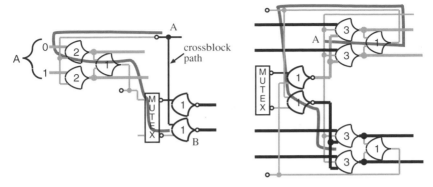

a. Transition to block timing issue   b. Transition to unblock timing issue

Figure 8.22   Timing issue for crossblocking signal.

from branch A to the 3 of 3 operators faster than the path indicated by the gray highlight.

8.5   CONCURRENT CROSSBAR NETWORK

8.5.1   The Arbitrated Crossbar Cell

The resources are now available to construct a concurrent crossbar. An arbitrated 2×2 crossbar cell where each of two input data paths can be steered to one of

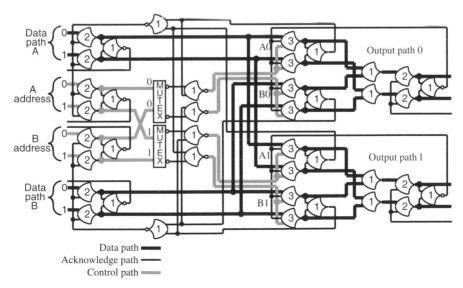

Figure 8.23   Arbitrated crossbar cell.

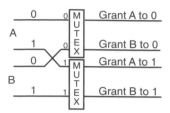

Figure 8.24 Dual-rail MUTEX.

two output data paths is shown in Figure 8.23. The 2 value input data paths, A and B, each represent a whole data path including the address variables. Each DATA wavefront will carry its own address with it. At each stage one 2 value address variable is stripped out of the data path to control the switch. The acknowledge to the 2 value input registration stage acknowledges the whole data path, including the stripped-out address variable.

The A and B address variables are arbitrated directly with a dual-rail MUTEX arbiter, as shown in Figure 8.24. The dual-rail MUTEX blocks an address and

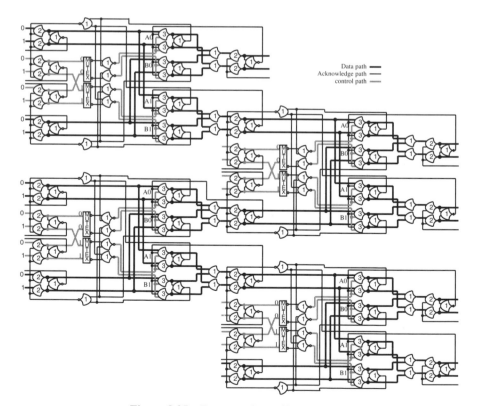

Figure 8.25 Four crossbar cells assembled.

wavefront only if the addresses conflict by addressing the same output path. If the two simultaneous wavefronts are addressed to different output paths, there is no blocking, and they flow with full concurrency. If A is directed to 0 and B is directed to 1, then both grants occur simultaneously.

A larger $N \times N$ crossbar network is assembled from this basic cell as in Figure 8.25. Wavefronts will flow through the network structure with maximal concurrency, and all flow conflict will be automatically and locally arbitrated.

8.5.2 2D Pipelining Arbitrated Control Variables

The crossbar structure and any other arbitrated structure can be 2D pipelined. The decisions of the arbiter manifests as a sequence of control variables, just as if they had emerged from a control path pipeline. Arbitration can be performed for the lead LSV or lead MSV, and the resulting control variables can be pipelined across the data path just as any other control variable.

8.6 EXERCISES

8.1. Design and simulate a 2D pipelined crossbar cell.

8.2. Design and simulate a 2D pipelined version of the interwoven control structure of Section 8.1.1.

8.3. Design and simulate a 2D pipelined serial bus.

CHAPTER 9

Multi-value Numeric Design

2NCL is an inherently multi-valued logic. Variables can be constructed from any number of values with any assigned meaning. These heterogeneous variables can be directly combined by appropriately combining their values. For instance, Section 4.4 of Chapter 4 presents a 2NCL expression adding a binary variable to a trinary variable producing a quaternary variable. This generality allows many conceptual conveniences in design specification. The question of this chapter is whether multi-value representation provides conveniences for expressing numeric processes.

9.1 NUMERIC REPRESENTATION

Multi-value variable representation allows the consideration of radices and encodings other than binary for numeric representation. This section will focus on cost of representation of numeric values, both for transmission and functional combination. The costs will be considered in terms of quantity of paths and operators to characterize physical resources and in terms of quantity of path transitions to characterize power and speed.

9.1.1 Resource Cost of Transmission

The total cost of transmitting a numeric representation is the cost of representing the values per digit times the cost of representing each digit. For the familiar electronic representation where a digit is a path and values are represented as discriminable voltage levels, values are clearly much more expensive than digits. In the multi-path environment, values and digits are both represented with paths. The cost to represent a digit and the cost to represent its values are identical, and there is no bias.

The cost in paths of representing a given number with M of N encoding is the number of values per digit times the number of digits to represent the number. Figure 9.1 shows the cost in paths of various encodings to represent numbers as the numeric magnitude increases. The standard binary is included as a reference.

Logically Determined Design: Clockless System Design with NULL Convention Logic^{TM}, by Karl M. Fant
ISBN 0-471-68478-3   Copyright © 2005 John Wiley & Sons, Inc.

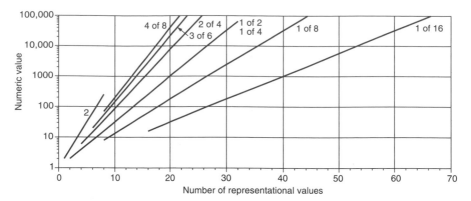

Figure 9.1   Cost in path resources.

The cost of discriminating two voltage levels is ignored, since both representations must bear this cost.

The 1 of N and $N/2$ of N encodings are considered. The $N/2$ of N encodings are more efficient in a specific respect than 1 of N encodings. For instance, a 1 of 4 encoding can represent four values per digit, while a 2 of 4 encoding can represent six values per digit. The thing to note at this point is that the $N/2$ of N encodings show a significant advantage and that the 1 of 2 and 1 of 4 encodings are the optimal 1 of N encodings.

9.1.2   Energy Cost of Transmission

The next question involves the energy of transmission. In the multi-path environment this can be quantified directly as how many paths switch to represent a number. In this the higher radices are superior because, although a radix 16 representation may require 16 paths per digit, it has fewer digits and only one path per digit switches.

Figure 9.2 shows the relative energy costs for various representations in terms of quantity of switches per number. The multi-path encoded digits switch twice per number because of the NULL transition between each DATA wavefront. Ignoring the clock signal, the binary representation switches 0.5 times per number representation because of the average switching behavior of binary representation. The figure shows that the encodings that use the most resources, and use each resource most sparingly, use the least energy.

The important observation from Figure 9.2 is that 1 of 4 encoding stands out as better than any of the $N/2$ of N encodings.

9.1.3   Combined Transmission Costs

Adding the two costs to get a figure of merit produces the chart in Figure 9.3. One can see that 1 of 4 is the optimal 1 of N encoding. It is superior to 1 of 2. Only 4 of 8 and 3 of 6 appear to be better.

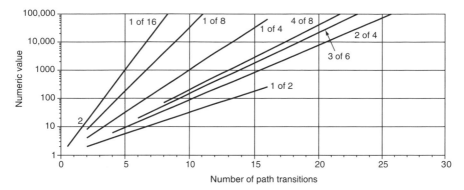

Figure 9.2 Switching costs of various representations.

9.1.4 Resource Cost of Combination

The next question is of the cost of combining numbers represented in the various encodings with arithmetic operations such as addition. The cost of combination is measured by taking the number of minterms defining a 2 input combinational function. Combining two 2 value digits, for instance, requires 4 minterms. Combining two 4 value digits requires 16 minterms. Figure 9.4 shows the cost in terms of quantity of minterms to combine increasingly large numbers for the various representations. Again, 1 of 4 stands out as an optimal representation only bested by 1 of 2.

9.1.5 Energy Cost of Combination

The last consideration is the energy used in combining numbers of various representations. This consideration is based on the number of minterm transitions in resolving the combination. The Boolean cost is 1.5 transitions per combination. There are 4 minterms per digit. The probability that the same minterm will be used consecutively and not switch is 0.25. The probability that a different minterm will be

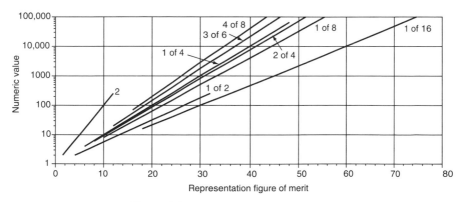

Figure 9.3 Combined transmission cost.

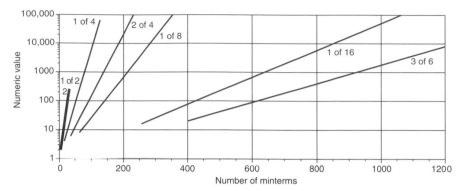

Figure 9.4   Resource cost of combination.

used and cause a transition is 0.75. If a different minterm transitions to TRUE, then the previously asserted minterm must transition to FALSE. Hence 1.5 transitions per digit combination. For all the other encodings exactly one minterm will transition to DATA then to NULL for two transitions per digit combination.

Figure 9.5 does not take into account the complexity of the minterm itself. For example, all of the 1 of N minterms will have two imputs but a 2 of 4 minterm will have four inputs and a 3 of 6 minterm will have six inputs. Also the fan-out of each input value for $N/2$ of N encodings will be higher than the fan-out for the 1 of N encodings.

Again, the encodings that use the most resources and that use each resource most sparingly use the least power.

9.1.6   Combined Cost for Numeric Combination

When the resource and energy costs of numeric combination are added in Figure 9.6, the result again shows that 1 of 4 is a most efficient encoding.

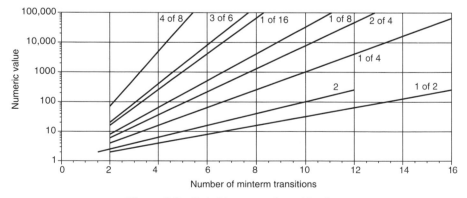

Figure 9.5   Switching cost of combination.

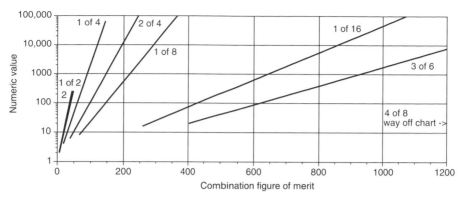

Figure 9.6 Combined cost of numeric combination.

9.1.7 Summary of Multi-path Numeric Representation

The bottom line is that 1 of 4 encoding seems to be a viable option for numeric representation with multi-path encoding. It is clearly superior to 1 of 2 encoding for transmission. The path resources for 1 of 2 and 1 of 4 are identical, and 1 of 4 requires half the switching energy. While 4 of 8 and 3 of 6 provide slightly better performances for transmission, their combinational performance is abysmal. Converting them to more efficient combinational codes is very expensive. The 2 of 4 code trails 1 of 4 in both categories. So the $N/2$ of N encodings are not in the competition. The 1 of $N > 4$ codes fall below 1 of 2 and 1 of 4 in both categories, so that leaves only 1 of 2 and 1 of 4 to consider.

While combining 1 of 4 variables is more costly in resources, it has, as shown in Figure 9.5, a considerable advantage in terms of energy and also speed (fewer stages of logic, fewer digits asserting a value, shorter addition carry chain). So it may be well worthwhile to develop, in the context of 2NCL logically determined systems, 1 of 4 or quaternary arithmetic functions. To explore this possibility, a quaternary

AND 2 bit binary encoding

		A			
		00	01	10	11
	00	00	00	00	00
B	01	00	01	00	01
	10	00	00	10	10
	11	00	01	10	11

AND Four rail encoding

		A			
		0	1	2	3
	0	0	0	0	0
B	1	0	1	0	1
	2	0	0	2	2
	3	0	1	2	3

OR 2 bit binary encoding

		A			
		00	01	10	11
	00	00	01	10	11
B	01	01	01	11	11
	10	10	11	10	11
	11	11	11	11	11

OR Four rail encoding

		A			
		0	1	2	3
	0	0	1	2	3
B	1	1	1	3	3
	2	2	3	2	3
	3	3	3	3	3

XOR 2 bit binary encoding

		A			
		00	01	10	11
	00	00	01	10	11
B	01	01	00	11	10
	10	10	11	00	01
	11	11	10	01	00

XOR Four rail encoding

		A			
		0	1	2	3
	0	0	1	2	3
B	1	1	0	3	2
	2	2	3	0	1
	3	3	2	1	0

NOT 2 bit binary encoding

	A			
	00	01	10	11
	11	10	01	00

NOT Four rail encoding

	A			
	0	1	2	3
	3	2	1	0

Figure 9.7 Derivation of quaternary logic functions.

SUM Four rail encoding

B\A	0	1	2	3
0	0	1	2	3
1	1	2	3	0
2	2	3	0	1
3	3	0	1	2

CARRY Four rail encoding

B\A	0	1	2	3
0	0	0	0	0
1	0	0	0	1
2	0	0	1	1
3	0	1	1	1

Figure 9.8   Quaternary addition.

ALU is presented and compared to a binary ALU with input and output conversions between 1 of 4 and 1 of 2 encodings.

9.2   A QUATERNARY ALU

In this section a quaternary ALU is presented with an adder, logical operators, and a 1 bit shift operator. The quaternary logic operators are derived in Figure 9.7 by

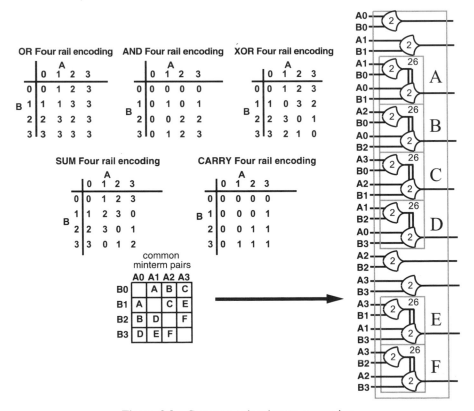

Figure 9.9   Common pair minterm expression.

considering the logical operation on two binary variables and mapping the results to one quaternary variable. The addition sum and carry of Figure 9.8 is a straightforward matter of radix 4 arithmetic.

In implementing multiple functions with a common input, it may be most efficient to express the minterm of the common inputs and then collect the various functions from the minterm. In this instance there is a commonality of mapping of the minterms that can be exploited. There are 6 pairs of minterms where each member of the pair generates a common result value for each function. The advantage is that the 16 minterms can be combined into 10 operators. So a common minterm can be expressed with 10 operators asserting 10 values of a single common minterm variable. The common pairs and the minterm expression are shown in Figure 9.9. The A terms are equal for all functions. The B terms are equal for all functions, and so on.

Figure 9.10 shows the logic functions using the common minterm expression and including the enabling control value for each function. Figure 9.11 shows the quaternary adder using the common minterm expression with its enabling control value.

There are two functions that use only input A: the NOT and the 1 bit shift. A 1 bit left or right shifter can be implemented directly in 4 value logic. Again, the function is derived from the effect of the shift on two binary variables. Figure 9.12 shows the derivation and the shift expression. The derivation of the NOT function and its expression are shown in Figure 9.13.

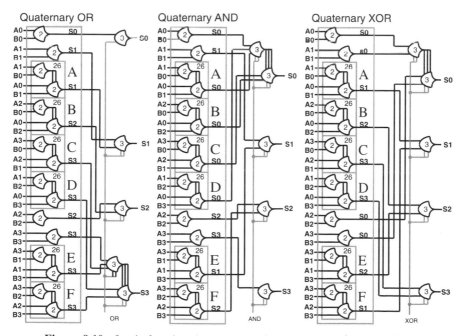

Figure 9.10   Logic functions implemented from common minterm variable.

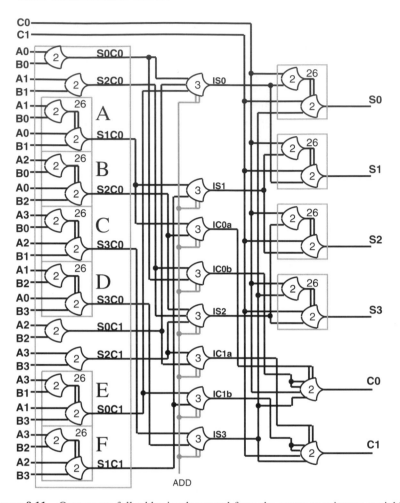

Figure 9.11   Quaternary full adder implemented from the common minterm variable.

The pieces are in place. In each expression defined above the control value that chooses the function and enables the function is combined into the expression in the first rank of logic of the function. In the case of the ADD and logic functions it is in the first rank of logic receiving the common minterm variable. This limits orphan paths of the input data paths. Unless a function is selected, no data path wavefront will pass through any operators of the function.

An ALU of any data path width can be assembled by stacking the above-defined digit functions across the data path and embedding them in an appropriate control structure, as shown in Figure 9.14. There is a 7 value command variable that selects the function for the ALU. The command variable is a wavefront, just like any other wavefront in the system. A command variable wavefront presents a single DATA

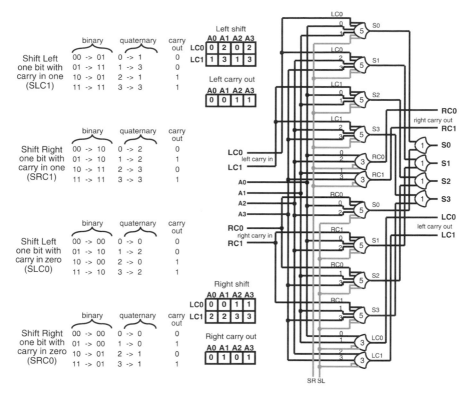

Shift Left one bit with carry in one (SLC1)

binary	quaternary	carry out
00 -> 01	0 -> 1	0
01 -> 11	1 -> 3	0
10 -> 01	2 -> 1	1
11 -> 11	3 -> 3	1

Left shift

	AO	A1	A2	A3
LC0	0	2	0	2
LC1	1	3	1	3

Left carry out

AO	A1	A2	A3
0	0	1	1

Shift Right one bit with carry in one (SRC1)

binary	quaternary	carry out
00 -> 10	0 -> 2	0
01 -> 10	1 -> 2	1
10 -> 11	2 -> 3	0
11 -> 11	3 -> 3	1

Shift Left one bit with carry in zero (SLC0)

binary	quaternary	carry out
00 -> 00	0 -> 0	0
01 -> 10	1 -> 2	0
10 -> 00	2 -> 0	1
11 -> 10	3 -> 2	1

Right shift

	AO	A1	A2	A3
LC0	0	0	1	1
LC1	2	2	3	3

Right carry out

AO	A1	A2	A3
0	1	0	1

Shift Right one bit with carry in zero (SRC0)

binary	quaternary	carry out
00 -> 00	0 -> 0	0
01 -> 00	1 -> 0	1
10 -> 01	2 -> 1	0
11 -> 01	3 -> 1	1

Figure 9.12 Derivation and expression for the quaternary shift.

value, enabling one function. When the function is complete, the command variable will present a NULL wavefront. When the NULL wavefront is complete, another command variable will be presented, selecting the next function. The input to the ALU is the data path wavefront and the command variable wavefront. Each Nth command variable wavefront operates on the Nth data path wavefront.

The ALU is a cycle with three input paths and three output paths. The output paths are well conditioned in that only one wavefront flows through one output

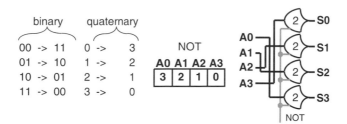

binary	quaternary
00 -> 11	0 -> 3
01 -> 10	1 -> 2
10 -> 01	2 -> 1
11 -> 00	3 -> 0

NOT

AO	A1	A2	A3
3	2	1	0

Figure 9.13 Derivation and expression for the quaternary NOT.

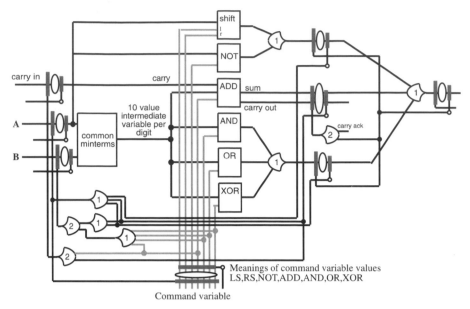

Figure 9.14   Quaternary ALU structure.

path for each command. There is no ambiguity about the generation of an acknowledge. There is, however, ambiguity about how the acknowledges are used. The input paths are not well conditioned in that some commands use three input paths (ADD), some commands use two input paths (AND, OR, XOR) and some commands use only one path (SHIFT, NOT). This is another situation where a completeness relationship varies with a controlling variable.

In this case the acknowledge paths to the input paths are conditioned by combining the command variable values with the acknowledge paths to the B and CARRY IN inputs. The command variable values enable the acknowledge paths, and hence enable the proper wavefronts, to flow through the proper input paths. All functions use the A input. The logical and ADD functions also use the B input and only the ADD function uses the CARRY IN input. Notice that the inversion of the acknowledges is before the merging of the command values. This puts the command variable, the input registration stages, and the acknowledges all in the same monotonic transition phase. Every cycle always uses the A input and the command input, so they are always acknowledged. This acknowledge is not conditioned by a command value.

9.3   A BINARY ALU

The next question is how the quaternary ALU compares with a binary ALU converting 4 value to 2 value, performing binary functions, and then converting back to 4

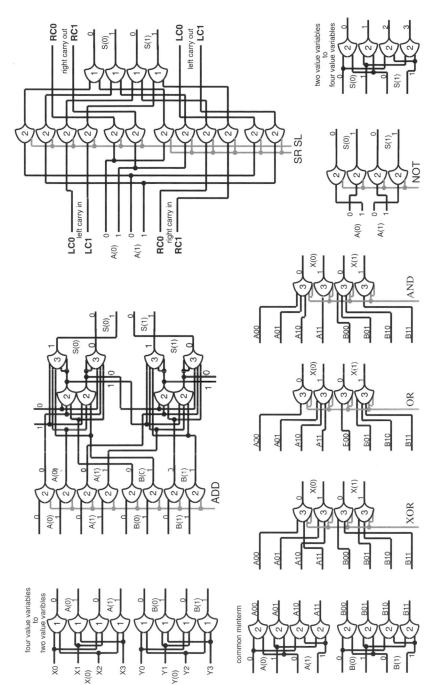

Figure 9.15 Binary ALU expressions.

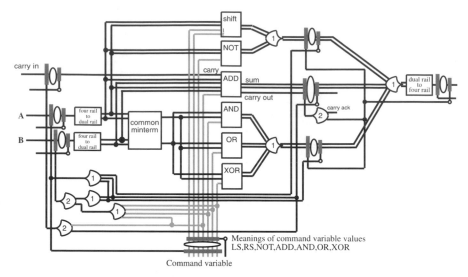

Figure 9.16 Binary ALU.

value. Figure 9.15 shows the function expressions for the binary ALU. There is the 4 value to 2 value conversion and the 2 value to 4 value conversion, and each function labeled with its command value name. The logic functions OR, AND, and XOR take advantage of a common minterm. The SHIFT, ADD, and NOT functions relate directly to the dual-rail inputs. The NOT function is almost identical for both the 4 value and the 2 value functions.

The basic structure of the binary ALU of Figure 9.16 is the same as the quaternary ALU except for the conversions between dual-rail and four-rail representations.

9.4 COMPARISON

The quaternary ALU is compared with the binary ALU with and without the data path conversions. The comparisons are in terms of operator count, propagation delay, and operator switching. The comparison is for the costs of processing one quaternary digit applied to a one digit ALU. The comparisons are tallied in Table 9.1. The binary ALU is presented with the conversion between four-rail and dual-rail, and it is presented without the conversion as if dual-rail data paths were presented directly to the ALU. The operator total is for the whole ALU. The switch and delay totals are just for the functions of the ALU, and each function includes the data path conversions and common minterm switches and delays.

For the two designs with a four-rail data path, the quaternary ALU wins on all counts. It is smaller and faster, and has much lower power. The quaternary ALU is even competitive with the binary ALU without the data path conversions. It is a little larger, but it is still faster and uses less power.

TABLE 9.1 Comparison of the Binary and Quaternary ALUs

	Binary ALU With Conversion			Quaternary ALU			Binary ALU Without Conversion		
	Operators	Switches	Delay	Operators	Switches	Delay	Operators	Switches	Delay
4 to 2	8	4	1						
2 to 4	4	1	1						
com min	8	1	1	10	1	1	8	1	1
ADD	16	13	7	14	5	3	16	8	5
OR	4	9	4	5	2.3	2.3	4	4	2
AND	4	9	4	5	2.3	2.3	4	4	2
XOR	4	9	4	5	2.2	2.2	4	4	2
NOT	4	7	3	4	1	1	4	2	1
SHIFT	16	11	4	16	4	2	16	6	2
Totals	68	58	26	59	16.8	12.8	56	28	14

9.5 SUMMARY

While the encoding analysis and the architecture comparison are somewhat cursory, they are sufficient to demonstrate that 1 of 4 encoding is an optimal numeric representation and that quaternary arithmetic with 1 of 4 encoding is a very viable option for logically determined system design.

9.6 EXERCISES

9.1. Discuss fours complement arithmetic, and explain how to deal with the sign digit.

9.2. Design and simulate a completely quaternary micro controller with fours complement arithmetic.

CHAPTER 10

The Shadow Model of Pipeline Behavior

The flow of wavefronts through a logically determined pipeline derives from the behavior of individual cycles and the shared path completeness synchronization among cycles. The flow of elements controlled by local interactions among the elements is an extraordinarily diverse phenomenon and has been studied using extraordinarily diverse models including queueing theory [3,62,38,37], traffic flow models [21], random walk models [14,12], industrial manufacturing production line models [18], and computer pipeline models [16,9,39,5,23]. Helbing [20] provides a comprehensive overview of the approaches to modeling self-driven many particle systems. None of these approaches, however, have generated an intuitively graspable model that illuminates the fundamental mechanism of wavefront flow through a logically determined pipeline structure. The shadow model fills this void.

10.1 PIPELINE STRUCTURE

The primary performance measure for a logically determined pipeline is throughput. The throughput of a cycle is the number of wavefronts that can propagate through a cycle per unit time. The throughput of a pipeline is the number of wavefronts that can propagate through the pipeline per unit time. The pipeline throughput is a result of the individual behaviors of its component cycles.

The first step is to define the behavioral components of the pipeline.

10.1.1 The Cycle Path and the Cycle Period

The cycle period is the sum of the delays around a cycle path as shown in Figure 10.1 For the examples of this chapter the 2 of 2 operator has a delay of three tics and the 1 of 2 operator with inverter has a delay of one tic. The cycle of Figure 10.1 has a period of 7 tics.

Logically Determined Design: Clockless System Design with NULL Convention Logic[TM], by Karl M. Fant
ISBN 0-471-68478-3   Copyright © 2005 John Wiley & Sons, Inc.

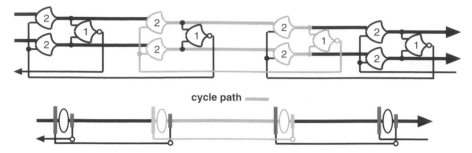

cycle path ▬▬

Figure 10.1 Cycle path.

10.1.2 The Wavefront Path: Forward Latency

The wavefront path is the data path, shown as gray in Figure 10.2, over which the wavefronts flow. The forward latency is the time that it takes for a wavefront to flow through the pipeline with no waits and is the sum of all delays along the wavefront path, which includes combinational expressions and registration operators.

10.1.3 The Bubble Path: Reverse Latency

The bubble path is the acknowledge path, shown as gray in Figure 10.3, over which the bubbles flow. The reverse latency is the time that it takes for a bubble to flow through the pipeline with no waits and is the sum of all delays along the bubble path, which includes registration operators, completion logic, and combinational logic in the acknowledge path.

10.2 THE PIPELINE SIMULATION MODEL

The space-time diagrams of the flow of successive wavefronts through successive pipeline cycles are generated by a Microsoft Excel [34] simulation (see Appendix C). Figure 10.4 is the baseline graph generated from the simulation showing the

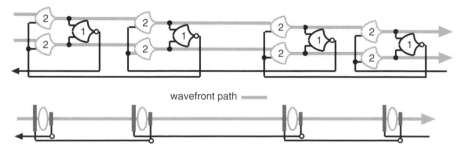

wavefront path ▬▬

Figure 10.2 Pipeline wavefront path or forward latency.

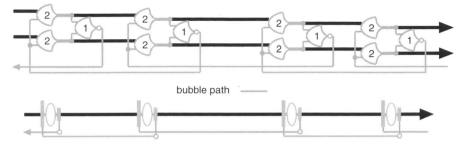

Figure 10.3   Pipeline bubble path or reverse latency.

space-time relationships among the successive wave front components propagating through the pipeline stages.

NULL and DATA wavefronts are not differentiated in the graph. In the context of performance it is irrelevant whether a wavefront is a DATA wavefront or a NULL wavefront. A wavefront is represented by a line flowing from upper left diagonally down toward lower right. It flows through the wavefront path defined by registration operator delays and data path propagation delays on the path. Bubbles are represented by the short disconnected lines between the wavefronts that flow from upper right to lower left. Bubbles flow through the acknowledge path defined by the completion delay and by the acknowledge propagation delay. Figure 10.5 shows the four delay components that determine the behavior of the pipeline.

Columns in the graph represent components of the wavefront path. A registration operator delay is represented by rx. A data path propagation delay is represented by

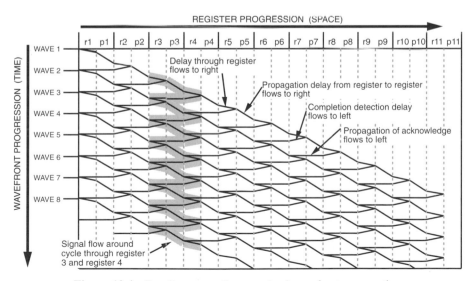

Figure 10.4   Baseline space-time graph of wavefront propagation.

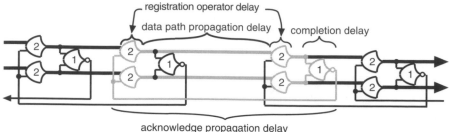

Figure 10.5   Delay components of pipeline behavior.

p*x*. Wavefronts flow progressively forward through r1, p1, r2, p2, r3, p3, ... Bubbles flow backward beginning at the right edge of an r*x* column, which represents the completion delay, and then flow backward through p*x* − 1 and r*x* − 1, which represents the acknowledge propagation delay. The left side of the graph, vertically labeled with successive wavefronts in time, is the input of the pipeline and the right side of the graph is the output of the pipeline. The pipeline modeled in the simulation is shown in Figure 10.6.

In the baseline graph of Figure 10.4 the delay of each component for each wavefront is identical, so the propagation of the wavefronts is very orderly. All signals are flowing in perfect synchrony and no signal has to wait on any other signal. The shaded region of the graph shows the cyclic signal flow, wavefront propagation forward and bubble propagation backward, around the cycle through propagation path 3.

10.3 DELAYS AFFECTING THROUGHPUT

The throughput of a logically determined pipeline is the number of wavefronts propagating through a cycle of the pipeline per time interval. If a given number of wavefronts take a longer time interval to propagate or if fewer wavefronts propagate in a given time interval, then the throughput is lower. If a given number of wavefronts propagate in a shorter interval or if more wavefronts propagate in a given time interval, then the throughput is higher.

A single slow signal event A (the delay of p5 for wave 3 is increased) is introduced in Figure 10.7, breaking the perfect synchrony of signal flow. The single

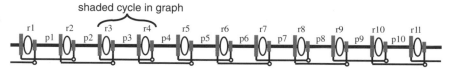

Figure 10.6   Pipeline modeled in simulation.

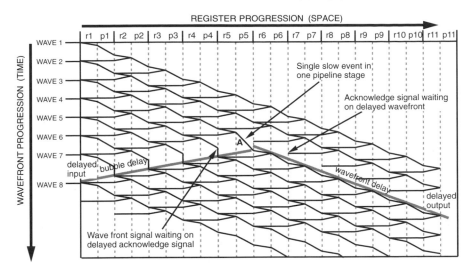

Figure 10.7   Effect of one slow signal.

delay event causes a delay that propagates forward through the pipeline as a wavefront delay causing bubbles to wait on the wavefront (a waiting bubble does not have a vertical line as the waiting wavefronts do). The single delay event also causes a delay that propagates backward through the pipeline along acknowledge signals as a bubble delay causing wavefronts to wait on the bubbles. These delays propagating to both the input and output of the pipeline cause a delay in time of succeeding wavefronts, reducing the throughput of the pipeline.

Typically the delays of interest will occur in combinational expressions in the data propagation path (prop x), but it does not matter where in the cycle an added delay occurs. The important factor is the cycle period. If a cycle is slow in any of its delay components, it will cause waits in other cycles. Figure 10.8 shows an increased delay A in the registration operator r5 for wave 3. Figure 10.9 shows an increased completion delay A after register 5 (r5) for wave 3. Figure 10.10 shows an increased acknowledge path delay A from the register 5 (r5) completion. In each case the behavior of the pipeline is identical. A delayed wavefront propagating forward causes bubbles to wait, and a delayed bubble propagating backward causes wavefronts to wait.

10.4   THE SHADOW MODEL

In Figure 10.11 a shadow is drawn radiating from the slow event A both forward and backward through the pipeline. The shadow represents the delay's domain of influence over other wavefronts flowing through the pipeline.

The bubble shadow projects backward from the delay. The lower boundary of the shadow is the bubble delay propagating backward causing wavefronts to wait.

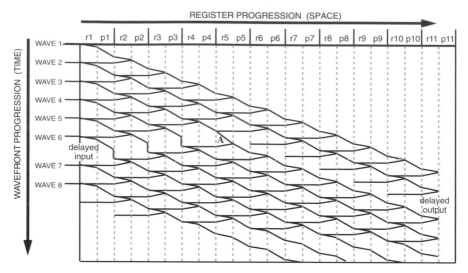

Figure 10.8   Increased delay in the register path.

The body of the shadow represents the region within which any delay causing a wait less than or equal to the wait caused by A will be absorbed by the wait caused by A. This is illustrated by event B in Figure 10.12. Delay B is shadowed by delay A.

The wavefront shadow projects forward from the delay. The lower boundary of the shadow is the wavefront delay propagating forward, causing bubbles to wait. The body of the shadow represents the region within which any delay causing a wait less

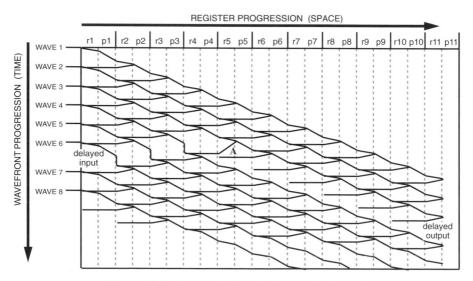

Figure 10.9   Increased delay in the completion path.

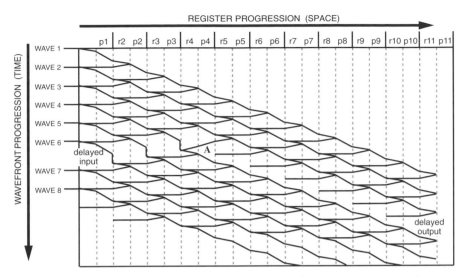

Figure 10.10   Increased delay in the acknowledge/request propagation path.

than or equal to the wait caused by A will be absorbed by the wait caused by A. This is illustrated by event C in Figure 10.12. Delay C is shadowed by delay A.

Smaller delays in the shadow of a larger delay simply distribute the larger delay over several smaller delays within the shadow of the larger delay and do not increase or decrease the total delay of succeeding wavefronts. So the smaller delays within

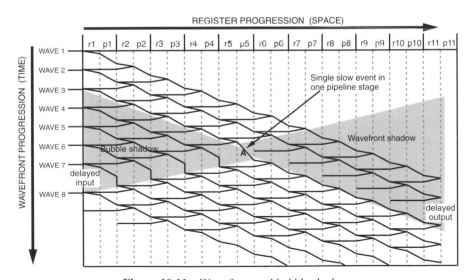

Figure 10.11   Wavefront and bubble shadows.

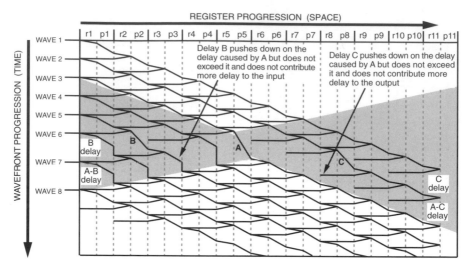

Figure 10.12 Smaller delays in the shadow of a larger delay.

the shadow do not affect the throughput of the pipeline. Neither delay B nor delay C affects the throughput of the pipeline.

Figure 10.13 shows how to understand what happens to the shadows of B and C. If there is no waiting, there is no shadow. The B delay casts a bubble shadow, causing wave 7 to wait at the input of the pipeline. But it has no effect on the wait of wave 8, which must wait on the bubble delay of A whether or not delay B

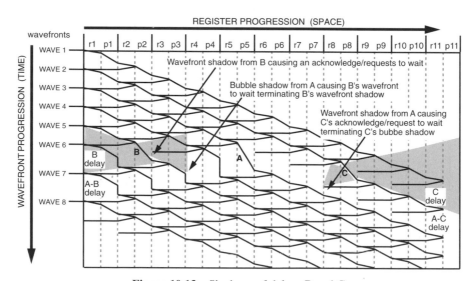

Figure 10.13 Shadows of delays B and C.

occurred. Its wavefront delay causes one bubble to wait and then encounters the bubble delay from A and itself has to wait. Once the B wavefront delay ceases causing a wait, it ceases casting a shadow.

The C delay casts a wavefront shadow, causing bubbles to wait at the output of the pipeline. Again, such a delay does not affect the delay of wave 3 caused by A. The bubble shadow of C immediately encounters the wavefront shadow of A, has to wait itself, and never causes anything to wait.

10.4.1 Shadowed Equal Delays

Figure 10.14 shows three equal delays that mutually shadow each other. The wavefront delay of B exactly matches the bubble delay of A, and neither causes any further waits after the encounter. The B wavefront shadow and the A bubble shadow exactly cancel. The wavefront delay of A exactly matches the bubble delay of C, and neither causes any further waits after the encounter. The A wavefront shadow and the C bubble shadow exactly cancel.

The wavefront shadow of C projects to the output, and the bubble shadow of B projects to the input causing a delay and affecting the through of the pipeline. Delay A has no effect on the through of the pipeline. There are three delay events in the pipeline, and they all shadow each other. The effect on the through of the pipeline is as if there were only one delay event in the pipeline.

10.4.2 Unshadowed Delays

Any unshadowed delay, however small, does have an effect on the throughput of the pipeline. Any delay above either shadow of A will cause A to be delayed in time.

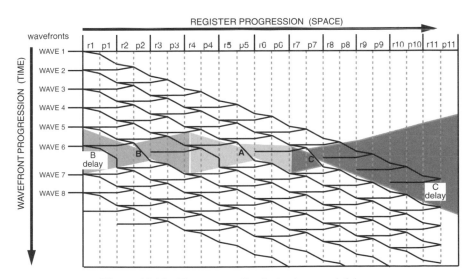

Figure 10.14 Three mutually shadowing equal delays.

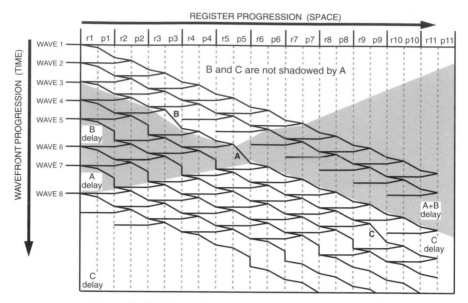

Figure 10.15 Three delays that do not shadow each other.

That delay will be added to the delay of A in delaying succeeding wavefronts in time, reducing the throughput of the pipeline. This is illustrated by event B in Figure 10.15. Delay B is not shadowed by delay A.

Any delay below either shadow will cause a delay in succeeding wavefronts in addition to the delay of A reducing the throughput of the pipeline. This is illustrated by event C in Figure 10.15. Delay C is not shadowed by delay A.

Figure 10.16 shows the shadows of B and Figure 10.17. show the shadows of C. Neither A, B, nor C shadow each other. All three delays are unshadowed, and all three delays affect the throughput of the pipeline.

10.4.3 Shadow Intersection

Delays that shadow each other compete to affect the throughput with the dominant shadow being effective and the dominated shadows being ineffective. Delays that do not shadow each other, however, do not compete but cooperate to form even more dominant shadows. The intersection of two shadows will shadow a delay that is the sum of the delays that cast the two shadows. In general, the intersection of N shadows from N nonshadowing delays will shadow a delay equal to the sum of the N delays.

Figure 10.18 shows how the bubble shadows of delays A, B, and C combine to shadow delays longer than any of the individual delays. Where the shadows of delay A and delay C intersect a delay must be larger than the sum of A and C to cause the delayed bubble of C to be later in time and affect the throughput of the pipeline.

Figure 10.19 shows the intersections of wavefront shadows of the three delays.

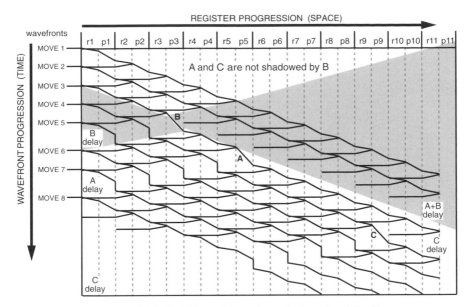

Figure 10.16 Bubble and wavefront shadows for delay B.

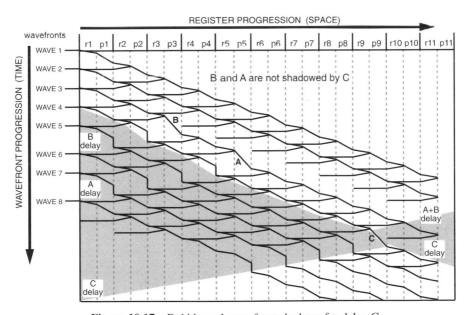

Figure 10.17 Bubble and wavefront shadows for delay C.

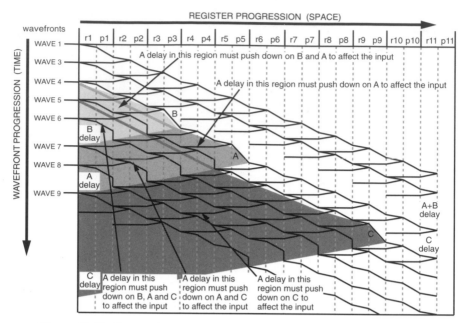

Figure 10.18   Intersections of bubble shadows for three nonshadowing delays.

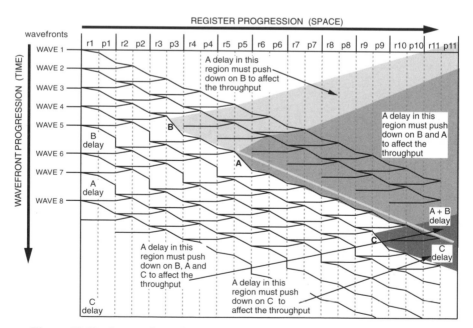

Figure 10.19   Intersections of wavefront shadows for three non-shadowing delays.

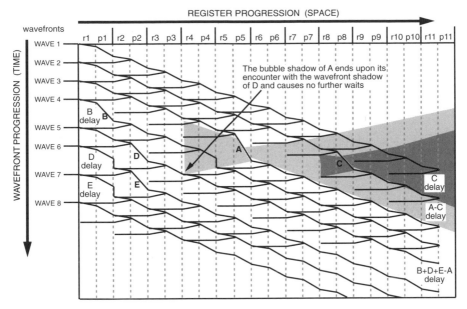

Figure 10.20 Shadows of delays A and C.

10.4.4 A More Complex Example

In Figure 10.20 shadowed and nonshadowed delays are combined in one example. Delays B, D, and E do not shadow each other. The shadows of B and D intersect to shadow A and A shadows C. The bubble shadow of C is terminated by the wavefront shadow of A. The influence of C never escapes the shadow of A. Although the B delay and the D delay are each shorter than the A delay, the intersections of their shadows shown in Figure 10.21 combine to shadow the A delay. The A delay occurs in an intersection region where a delay must be greater than the sum of the delays of B and D and E. The delay must push down the waits of B, D, and E to affect the throughput. A pushes down on the waits of B and D but does not push down on the wait of E and is thus shadowed by B, D, and E.

10.5 THE VALUE OF THE SHADOW MODEL

Pipeline behavior had been considered too complex and dynamic for human intuition to deal with. While shadow intersections can get quite complex (the behavior range of a pipeline is the combination of all the possible interactions among shadows), the shadow model allows one to eliminate a large amount of pipeline behavior as irrelevant to throughput, to focus on what is relevant, and to characterize collective behavior intuitively with a clear understanding of the mechanism underlying the behavior. Several known observations about logically determined pipeline

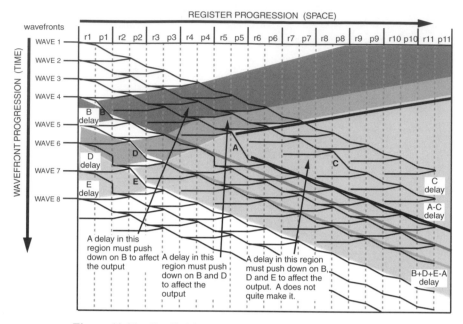

Figure 10.21   Small delays combining to shadow a larger delay.

behavior discovered by dynamic simulation can be understood directly in terms of the shadow model.

10.5.1   The Consistently Slow Cycle

A single consistently slowest cycle will dominate and determine the throughput of a pipeline [23]. Consider the pipeline of Figure 10.22 with a long delay in p5. The long delay of that cycle will continually send successive shadows through the pipeline, shadowing every other faster cycle in the pipeline which must wait on the delay shadows cast by the slow cycle. The throughput of the pipeline will be the throughput of that slowest cycle. No other faster cycle in the pipeline will affect the throughput of the pipeline. A pipeline is only as fast as its slowest cycle.

10.5.2   The Occasional Slow Cycle

Consider a pipeline composed of cycles with each cycle exhibiting a distribution of throughput or delay behavior uncorrelated with the other cycles. This throughput variability may accrue from the nature of a combinational expression in the cycle, for instance. Each cycle in isolation will have an average case throughput. Every time any cycle has a long delay, the shadow of this long delay will project through the pipeline shadowing all faster cycles in the pipeline. This causes wavefronts to be delayed in time and the throughput of the pipeline to decrease.

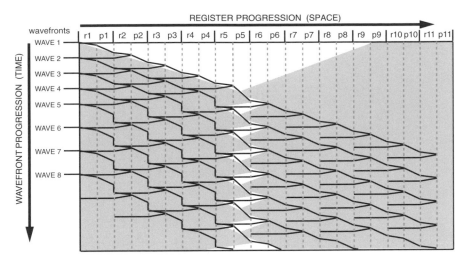

Figure 10.22   Pipeline with one consistently slow cycle.

A slow cycle period will almost always cast an effective shadow, decreasing the throughput of a pipeline. A fast cycle period will almost always be shadowed and have no effect on the throughput of a pipeline. The only time that fast cycle periods can increase the throughput of the pipeline is on rare occasions when there are no slow cycles casting shadows through the pipeline. The result is that the throughput of the pipeline as a whole will tend toward the worst-case throughput of its individual cycles [23,17].

10.6   EXERCISES

10.1. Describe how shadows behave through pipeline fan-out and through pipeline fan-in.

10.2. Describe the behavior of a system at startup as the first data wavefronts encounter slow cycles inside the system.

CHAPTER 11

Pipeline Buffering

A buffer is a fast cycle added to a pipeline. It is typically, but not necessarily, nonfunctional except for its buffering duties. There are three reasons for adding buffer cycles to a pipeline. The first is to enhance the throughput of a pipeline composed of variable throughput cycles. The second is to synchronize variable throughput behavior with constant throughput behavior. These first two reasons are the subject of this chapter.

The third reason for adding buffer cycles to a pipeline is to optimize the throughput of a structurally bounded pipeline, which is the subject of the next three chapters.

11.1 ENHANCING THROUGHPUT

Adding buffer cycles enhances the throughput of a pipeline composed of variable throughput cycles [23]. As was pointed out in the last chapter, the throughput of a pipeline tends to the worst-case throughput of its component cycles. The idea is to bring the throughput of the pipeline closer to the average case throughput of its component cycles. A buffer cycle in this instance is a cycle whose period is always equal to or faster than the fastest period of the variable throughput cycles. In this circumstance there is no advantage to a buffer cycle being faster than the fastest period of the variable throughput cycles. Faster buffer cycles can enhance pipeline latency, but they will not enhance pipeline throughput because they will always be shadowed by the variable throughput cycles.

Buffer cycles make variable throughput cycles farther apart without introducing more slow cycle delays. The effect of this is to increase shadow coverage between the variable throughput cycles, which makes it more likely that long cycle delays will shadow each other and lessen their overall effect on the pipeline throughput. Recall from Section 10.4.1 of Chapter 10 that if three long delays shadow each other the effect on the throughput is of only one long delay.

Figure 11.1 shows time-space diagrams with cycles represented in vertical columns and wavefront flow represented in horizontal rows. There is one long delay represented by the black cell in the center. The delay occurs on wavefront 0 in

Logically Determined Design: Clockless System Design with NULL Convention Logic^{TM}, by Karl M. Fant
ISBN 0-471-68478-3 Copyright © 2005 John Wiley & Sons, Inc.

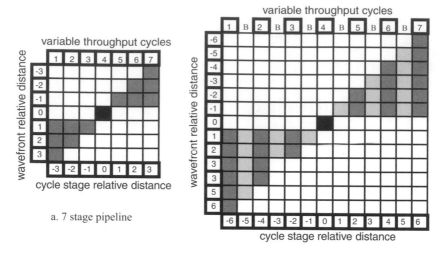

a. 7 stage pipeline

b. 13 stage pipeline

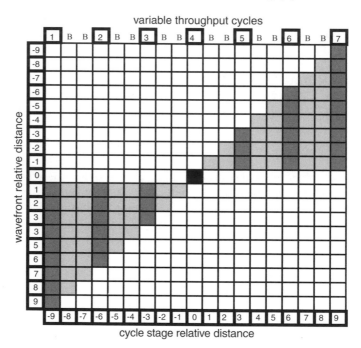

c. 19 stage pipeline

Figure 11.1   Shadow coverage with intermingled buffers.

the middle of the pipeline. The wavefronts and cycles are labeled relative to the long delay. Wavefronts that preceded the wavefront with the delay are labeled with negative numbers. Wavefronts after the long delay wavefront are labeled with positive numbers. Similarly cycles preceding the delay cycle are labeled with negative numbers and cycles after the delay cycle are labeled with positive numbers.

Each square represents a wavefront in a cycle. The shaded squares represent the shadows projecting from the long delay. Darker squares represent shadowed variable delay cycles. Lighter squares represent shadowed buffer cycles. The projection leftward and downward is the bubble shadow propagating backward through the pipeline and affecting wavefronts succeeding the delay wavefront. The projection rightward and upward is the wavefront shadow propagating forward through the pipeline and affecting the bubbles of wavefronts preceding the delay wavefront.

Figure 11.1a shows the initial pipeline with 7 variable throughput cycles. The long delay of cycle 4 will shadow one wavefront in cycles 5 and 3, two wavefronts in cycles 6 and 2, and three wavefronts in cycles 7 and 1. If any of these shadowed wavefronts were a long delay, then the long delays would shadow each other.

Figure 11.1b shows a 13 cycle pipeline with a buffer cycle between each variable throughput cycle. Since the variable cycles are now farther apart, the delay shadow covers more wavefronts in each variable cycle. The long delay shadow now covers two wavefronts in the variable cycles 3 and 5 so that if either of the two wavefronts were long delays they would be shadowed. It covers 4 wavefronts in cycles 2 and 6 and covers 6 wavefronts in cycles 1 and 7. With the single buffer cycles between each variable cycle the shadow coverage is doubled, making it twice as likely that the long delay in the middle cycle will shadow or be shadowed by a long delay in another cycle. This is true for all cycles. It is twice as likely that any long delay anywhere in the pipeline will be shadowed by another long delay somewhere else in the pipeline, lessening the effect of long delays on the throughput of the pipeline.

Figure 11.1c shows a 19 cycle pipeline with two buffer cycles between each variable cycle. The shadow coverage of the long delay is now three times larger for each variable cycle than it was for the pipeline with no buffers, making it three times as likely that long cycle delays will shadow each other.

This increase of shadow coverage is the mechanism of buffering that increases pipeline throughput. As more buffers are added, the performance of the pipeline will asymptotically approach the average behavior of individual cycles. The shadow model also exhibits the exponential decrease of efficacy with the increase of buffers between variable cycles as observed in simulation studies [23]:

One buffer provides 100% increase in shadow coverage.

A second buffer provides 50% additional increase in shadow coverage.

A third buffer provides 33.3% additional increase in shadow coverage.

A fourth buffer provides a 25% additional increase in shadow coverage, and so on.

11.1.1 Buffer Structuring for Throughput

Is there an optimal placement of buffer cycles among variable throughput cycles? The answer has to be the configuration that places the maximum collective distance between variable cycles and maximizes the shadow coverage among them. It has been shown by simulation that the inverted bowl configuration, placing all the buffers in the middle of the pipelines, is superior to evenly distributing buffers among variable throughput cycles [62]. Why this is so can be seen in the table of Figure 11.2 where the shadow coverage for three configurations of a pipeline with 6 variable cycles and 5 buffer cycles is tallied.

In the table of Figure 11.2 A, B, C, D, E, and F are variable throughput cycles and b represents a buffer cycle. The vertical columns show the shadow coverage of each variable cycle for each other variable cycle. In the table of Figure 11.2*a*, variable cycle A covers 2 wavefronts of cycle B, 4 wavefronts of cycle C, and so on. Cycle B covers 2 wavefronts of cycle A, 2 wavefronts of cycle C, and 4 wavefronts of cycle D. In the table of Figure 11.2*b*, A covers 1 wavefront of cycle C, 2 wavefronts of cycle C, 8 wavefronts of cycle D, and so on. It can be seen that the total shadow coverage obtained from the 5 buffers is larger for the case of the buffers concentrated in the middle of the pipeline (160 for center grouped, 140 for distributed)

pipeline cycle	1	2	3	4	5	6	7	8	9	10	11
pipeline configuration	A	b	B	b	C	b	D	b	E	b	F
Shadow coverage for cycle A	0		2		4		6		8		10
Shadow coverage for cycle B	2		0		2		4		6		8
Shadow coverage for cycle C	4		2		0		2		4		6
Shadow coverage for cycle D	6		4		2		0		2		4
Shadow coverage for cycle E	8		6		4		2		0		2
Shadow coverage for cycle F	10		8		6		4		2		0
total shadow coverage	30		22		18		18		22		30 = 140

a. Distributed buffers

pipeline cycle	1	2	3	4	5	6	7	8	9	10	11
pipeline configuration	A	B	C	b	b	b	b	b	D	E	F
Shadow coverage for cycle A	0	1	2						8	9	10
Shadow coverage for cycle B	1	0	1						7	8	9
Shadow coverage for cycle C	2	1	0						6	7	8
Shadow coverage for cycle D	8	7	6						0	1	2
Shadow coverage for cycle E	9	8	7						1	0	1
Shadow coverage for cycle F	10	9	8						2	1	0
total shadow coverage	30	26	24						24	26	30 = 160

b. Center grouped buffers

pipeline cycle	1	2	3	4	5	6	7	8	9	10	11
pipeline configuration	b	b	A	B	C	D	E	F	b	b	b
Shadow coverage for cycle A			0	1	2	3	4	5			
Shadow coverage for cycle B			1	0	1	2	3	4			
Shadow coverage for cycle C			2	1	0	1	2	3			
Shadow coverage for cycle D			3	2	1	0	1	2			
Shadow coverage for cycle E			4	3	2	1	0	1			
Shadow coverage for cycle F			5	4	3	2	1	0			
total shadow coverage			15	11	9	9	11	15			=70

c. Edge grouped buffers

Figure 11.2   Shadow coverages for three buffer configurations.

and that more long delays will shadow each other and reduce their effect on the throughput.

It should be obvious to the reader that placing the buffers at the beginning or end of the pipeline leaving the variable cycles all grouped together, as shown in Figure 11.2c, will have no effect at all on the throughput of the pipeline. This configuration is equivalent to no buffering at all, and the shadow coverage is only 70.

11.1.2 Correlated Variable Cycle Behavior

A pipeline with correlated delay behavior may deliver higher throughput than one with uncorrelated delay behavior [3]. Since the delay of a cycle is typically data dependant and the distribution of the data is generally not controllable, it is unlikely that the relationships of the long delays can be managed. But, if they by some chance can be managed, then they can be placed in relationships to maximize mutual shadowing.

This is the case for a pipeline in which for each waverfront the delay is the same in all cycles; a slow waverfront is slow in all cycles and a fast waverfront is fast in all cycles. For any consecutive long delay wavefronts, the slow cycles of the first wavefront will shadow all but the last cycle delay of the long delays of the second wavefront, greatly reducing the overall effect of long delays on the throughput. In Figure 11.3 wavefront 3 and wavefront 5 have equally long delays through all cycles of the pipeline. Wavefronts 4, 6, 7, and 8 all have fast cycles. Wavefront 3 effectively shadows almost all of the delays of wavefront 5. Since wavefront data 3 is two wavefronts away from wavefront data 5, only the delays from the last

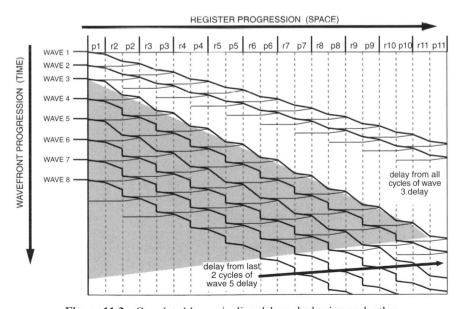

Figure 11.3   Correlated long pipeline delays shadowing each other.

two cycles of wavefront 5 are unshadowed and have an effect on the throughput. The bubble shadow of the last cycle of wave 3 is shown not shadowing the last two cycles of wave 5. All the other delays of wavefront 5 are shadowed by the delays of wavefront 3. Once the initial price is paid for the first unshadowed slow wavefront, successive slow wavefronts will largely shadow each other.

11.1.3   Summary of Throughput Buffering

Buffering works by increasing the pipeline distance between variable delay cycles and allowing long delays more opportunity to shadow each other lessening their affect on the throughput. Buffering to improve throughput is only effective when there is variability of delay behavior among the cycles. If the delay behavior of all cycles of a pipeline is constant, then buffering will have no effect on the throughput.

Whether or not two long delays shadow each other depends on the separation slope between them. The separation slope is how many wavefronts apart they are over how many cycles apart they are. If this ratio is less than 1.0, then they shadow each other. If it is greater than or equal to 1.0, then they do not shadow each other.

The farther apart long delays are in terms of wavefronts, the less likely they will shadow. The farther apart long delays are in terms of cycles, the more likely they will shadow. If long delays are rare, they will likely affect the throughput and not shadow, so buffering will have little effect. As they become more frequent, they will begin shadowing and lessen their collective effect on the throughput, so buffering will have a significant effect.

11.2   BUFFERING FOR CONSTANT RATE THROUGHPUT

How can the variable delay behavior of a logically determined pipeline be buffered to interact with constant throughput behavior? The answer can be understood in terms of a competition of intersecting shadows.

Figure 11.4 shows the baseline simulation configuration of a 10 cycle pipeline. Cycle 1 is the input cycle for the pipeline and cycle 10 is the output cycle for the pipeline, both of which must maintain a constant cycle period. They might represent two clocked interfaces, for instance. The cycle period for both cycles must be identical, but they need not be in phase. If the constant cycle periods are not identical, the pipeline between them will eventually either fill up and fail to provide a bubble for the input cycle or empty and fail to provide a wavefront for the output cycle. Cycle 5 is the variable delay behavior cycle. This cycle can represent a pipeline of any length with a variable delay behavior profile. Cycles 2, 3, 4, 6, 7, 8, and 9 are the buffer cycles. Variable delays will be inserted into cycle 5 to be buffered to constant delays into cycles 1 and 10. The periods of all the buffer cycles, and also initially that of cycle 5, are 3 tics faster than the input and output cycles.

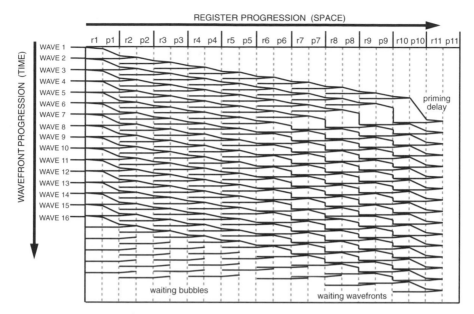

Figure 11.4   Baseline rate equalization buffer.

The basic problem is to ensure that the input and output cycles never have to wait. There must always be a bubble waiting to enter the input cycle, and there must always be a wavefront waiting to enter the output cycle. The basic buffering strategy is to keep a population of bubbles waiting to flow through the input cycle and to keep a population of wavefronts waiting to flow through the output cycle. These populations will wax and wane with the varying delay behavior of the variable delay cycle, but neither population must ever be depleted.

To effect the buffering a pipeline must be primed with a population of waiting wavefronts. Normally this would be accomplished by just initializing data wavefronts in the pipeline. The simulation does not support initialization of wavefronts, so the pipeline is primed with a long output delay at the beginning. The delay causes wavefronts that flowed through the input to back up in the pipeline as shown in Figure 11.4. After wave 5 passes through cycle 6, the pipeline is primed for buffering with a population of waiting bubbles and a population of waiting wavefronts.

11.2.1   The Buffering Behavior

Waits are caused by shadows projected from the input and output cycles. To keep the populations of bubbles and wavefronts waiting, the input and output cycle must project dominating shadows into the pipeline. This means that the periods of the input and output cycles must be greater than the periods of the buffer cycles. If the delays of the buffer cycles were greater, they would project dominant shadows. These shadows would cause the input and output cycles to wait, which is precisely what must not happen. So the buffer cycles must be strictly faster than the input and output cycles. To this end the buffer cycle periods of the simulation are 3 tics faster than the input and output cycles.

Wavefront shadows project forward into the pipeline from the input cycle causing bubbles to wait. Bubble shadows project backward into the pipeline from the output cycle causing wavefronts to wait. Since the input and output cycle delays are identical and greater than all the other delays in the pipeline, their shadows encounter no competition until they meet in the middle of the pipeline. By shadowing each other, the delays exactly cancel. The projected shadows are shown in Figure 11.5. The shadows are continually projected with every wavefront, but for purposes of illustration a single shadow from each end canceling in the middle is shown in the upper part of the diagram. Three consecutive intersecting shadows from each end are shown in the lower part of the diagram.

Each shadow intersection accumulates delay isolation for the input and output cycles. With each intersection it takes a longer delay to have an effect on the projecting cycle. This is explained in Section 10.4.3 of Chapter 10. Since the internal cycles are 3 tics faster than the input and output cycles, the projected shadows cause the bubbles and wavefronts in each cycle to wait 3 tics as they propagate through the pipeline.

Consider the bubble shadows projected from cycle 10. Cycle 9 is covered by only one shadow from cycle 10. Cycle 8 is covered by 2 shadows. Cycle 7 is covered by 3

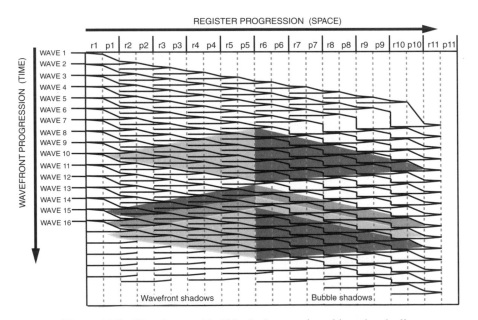

Figure 11.5 Wavefront and bubble shadows projected into the pipeline.

shadows, and so on. It would take 4 tics of additional delay in cycle 9 to affect the delay of cycle 10. For cycle 8 it would require 7 tics of additional delay to overcome the 3 tic wait of cycle 8 as well as the 3 tic wait of cycle 9 to affect the delay of cycle 10. Cycle 7 would require 10 tics of additional delay to affect cycle 10, and so on. These are the effects of the intersecting shadows. Cycle 5 sees a cascade of four 3 tic waits and, with its own 3 tic wait, would have to exhibit an additional delay of 16 tics to affect the delay of cycle 10.

The stage is now set to buffer the variable delays of cycle 5 and to clearly understand the mechanism and its limitations. Intersecting shadows projected from cycle 5 continually compete with intersecting shadows projected from the input and output cycles. If the input and output shadows always win, then the input and output cycles will never be delayed. If cycle 5 shadows ever win, then there will be a delay in cycle 1 and cycle 10.

The score of this competition is kept by the waiting wavefronts and bubbles in the pipeline. If, during a cycle 5 delay, all the wavefronts ahead of cycle 5 in the pipeline are emptied and the delayed wavefront from cycle 5 does not get to cycle 10 in time, then there will not be a wavefront to serve the next request of cycle 10. The wavefront shadows projected from cycle 5 overcome the bubble shadows projected from cycle10. If, during a cycle 5 delay, wavefronts flowing into the pipeline build up behind cycle 5, all the bubbles are emptied out, and the delayed bubble from cycle 5 does not get to cycle 1 in time, then there will not be a bubble to serve the next request of cycle 1. The bubble shadows projected from cycle 5 overcome the wavefront shadows projected from cycle 1. There must be enough wavefronts and bubbles with a sufficient budget of waits to accommodate the worst case behavior of the variable delays of cycle 5.

One way to understand the behavior is in terms of debits and credits to a wait account. The account is initialized with a deposit of waits. This initial deposit of waits depends on the number of buffer cycles, the number of wavefronts initialized in the pipeline, the period of the buffer cycles, and the period of the input and output cycles. The amount of wait maintained by each buffer cycle is the input–output cycle period minus the buffer cycle period. Unlike buffering for throughput where there is no advantage to making a buffer cycle faster than the slowest cycle of the variable cycles, here there is an advantage to making the buffer cycles as fast as possible. The faster a buffer cycle is in relation to the input–output cycles, the more wait it can maintain. This is illustrated in Figure 11.6, which is configured of buffer cycles whose period is 5 tics less than the input–output cycle periods. In this case cycle 5 must exhibit a 26 tic delay to affect the output cycles.

If the pipeline buffering is balanced in terms of the number of waiting wavefronts and waiting bubbles, then the behavior of the input and output buffering should be symmetric and the buffering behavior can be characterized just in terms of the waiting wavefronts. The number of wavefront cycles initialized times the amount of wait maintained by each cycle will determine the initial wait account for the pipeline. The baseline simulation is 4 wavefronts times 3 tics per wavefront for an initial account of 12 tics of wait.

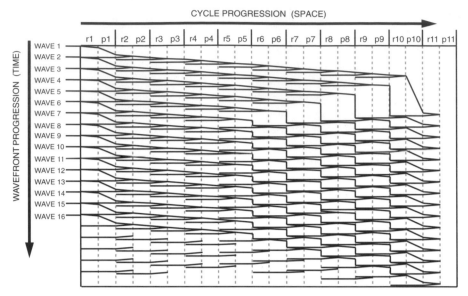

Figure 11.6   Faster buffer cycles maintaining more wait.

11.2.2   The Competition

Figure 11.7 shows a series of cycle 5 wavefronts that illustrates the shadow competition and the wait account behavior. Cycle 5 is represented by p5 in the chart. The first long delay of wavefront 7 in cycle 5 projects a wait shadow of 12 tics against the input–output cycle shadows. This projected wait uses up all the bubble waits of cycles 2, 3, and 4 and all the wavefront waits of cycles 6, 7, 8, and 9, it just fails to overcome the shadows of the end cycles. The transaction value of this delay is −12. Twelve tics of waits were withdrawn from the account.

With succeeding fast periods of cycle 5, bubbles and wavefronts rush through and replenish the wait account. The period of these wavefronts in cycle 5 are 3 tics shorter than the input–output cycles, and the transaction value for these cycles is +3. As each wavefront encounters the shadow from the output cycle and is caused to wait, it redeposits 3 tics of wait in the account. The bubble shadow of cycle 10 can be seen re-extending as wave 8 has to wait in cycles 9, wave 9 has to wait in cycle 8, wave 10 has to wait in cycle 7, and wave 11 has to wait in cycle 6. Finally, the shadows reach cycle 5, and although cycle 5 still has a fast period, it is now shadowed and has to wait. Continued fast behavior of cycle 5 will not increase the wait account nor in any way affect the behavior of the pipeline. The wait account is full again, and the shadows of the input–output cycles are re-extended to the center of the pipeline and mutually shadow each other.

In the end, a delay in wavefront 15 in cycle 5 causes a cycle period 15 tics longer than the end cycle period, which projects a wait shadow of 15 tics with a transaction of −15 against the refilled wait account. This uses up all 12 wait tics, overdrawing

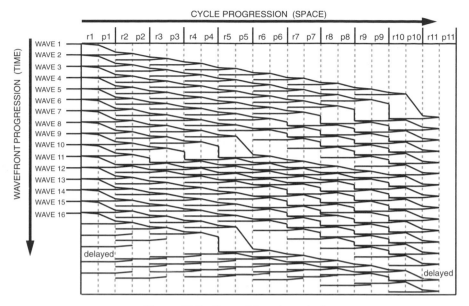

Figure 11.7 Long delays in cycle 5.

the account, and causes a 3 tic extension of the input–output cycle periods, delaying them in time by 3 tics.

11.2.3 The Battle of Intersecting Shadows

Consecutive delay shadows from cycle 5 can intersect and gang up on the intersecting shadows of the input–output cycles. This intersecting and ganging up is manifested fested as consecutive negative transactions. Figure 11.8 shows several consecutive delays in cycle 5, each of which causes a period of 6 tics greater than the input–output cycle periods projecting a 6 tic wait shadow and a transaction to the account of −6. No delay singly casts a sufficient shadow to reach an end cycle, but the shadows accumulate by intersection. They finally overdraw the wait account and overcome the input–output cycle shadows, causing delays in the input–output cycles.

Notice that once the wait account is overdrawn and a delay is caused in the input–output cycles, the continued −3 transactions do not increase the delay of the input–output cycles. There is no further increasing deficit applied to the input–output cycles. When the wait account is empty, delays from cycle 5 simply pass through the input–output cycles. When positive transactions occur, the wait account begins refilling immediately.

11.2.4 The Standoff

While negative value transactions withdraw from the wait account until it is empty and positive value transactions deposit to the wait account up to its limit; 0 value

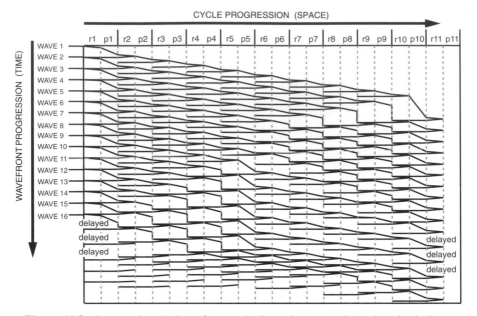

Figure 11.8   Intersecting shadows from cycle 5 ganging up on the end cycle shadows.

transactions do not affect the account at all. Figure 11.9 shows a sequence of delays in cycle 5 with transaction values of 0. There is an initial long delay in wavefront 7 of cycle 5 with a transaction value of -12 that empties the wait account. It does not, however, cause any delay in the input–output cycles. This wavefront is followed by a sequence of wavefronts that passes through cycle 5 with the same period as the input–output cycle periods, giving them a transaction value of 0. This series of wavefronts cast a shadow with the same wait value as the input–output cycles cast. These shadows meet and mutually shadow with no net effect on the behavior of the pipeline. The wait account does not change. Finally, in wavefront 14 in cycle 5, a delay causing a period 3 tics longer than the input–output cycles presents a -3 transaction to the depleted wait account and causes a delay in the input–output cycles.

11.2.5   Summary of Buffering for Constant Rate Throughput

If the input–output cycles must continually exhibit constant cycle periods, the wait account must never be overdrawn, and the input–output shadows must never be overcome. This depends on the behavior profile of the variable cycles in the pipeline and the capacity of the buffering wait account. The behavior profile must not only specify the statistical distribution of delays in the variable cycles but also the possibilities of consecutive delays with negative transaction values that can accumulate to deplete the wait account.

CYCLE PROGRESSION (SPACE)

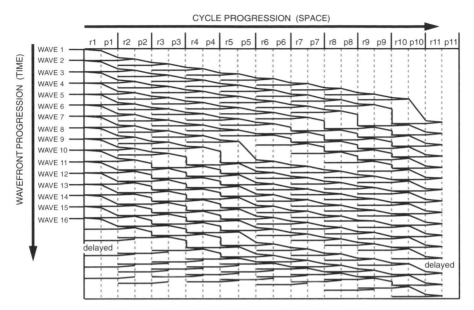

Figure 11.9 Delays with transactions values of 0.

The drill, in a nutshell, is to ensure that the input–output cycles cast shadows into the variable rate pipeline that are never challenged and overcome by any shadows cast by the variable rate pipeline itself. The only way to guarantee this is to ensure that no wavefronts in the variable rate pipeline have a cycle period greater than the input–output cycles.

11.3 SUMMARY OF BUFFERING

The basic idea underlying throughput buffering is to get the variable cycles as far apart as possible to maximize the shadow coverage of each cycle by the other cycles. The basic idea underlying constant throughput buffering is to keep a population of wavefronts and bubbles presented to the constant rate cycles so that they never have to wait. Both of these buffering behaviors can be understood in terms of shadows.

Both buffering tasks are accomplished by adding fast cycles to a pipeline. In the case of throughput buffering, there is no advantage to making the buffer cycles any faster than the fastest variable cycle. In the case of constant rate buffering, the faster the buffer cycles are the better.

If there is no variable delay behavior of the cycles of a pipeline, then throughput buffering can have no effect in either case. Constant rate buffering, on the other hand, always requires a certain amount of buffering. There must be a population of initialized wavefronts and the population must be maintained.

While adding buffers can increase the throughput of a pipeline, the buffers will never decrease the throughput. One can add all the buffers one likes to a pipeline and still not reduce the throughput. This will just increase the latency.

11.4 EXERCISES

11.1. Can making the NULL wavefront period of each cycle shorter than the DATA wavefront period of each cycle increase the throughput of a pipeline? Explain this in terms of shadows.

11.2. Define a procedure to determine the optimal buffering for a pipeline with given profiles of cycle behavior.

CHAPTER 12

Ring Behavior

A pipeline can be closed by connecting the inputs to the outputs, as shown in Figure 12.1 to form a pipeline ring. The behavior of a ring is determined by the closed nature of its structure. There have been previous characterizations of ring behavior [17,61]. This characterization does not differ from the literature in substance but differs in point of view which allows ring behavior to be much more intuitively and easily understood.

12.1 THE PIPELINE RING

The throughput performance of the ring is the number of wavefronts flowing through any cycle per unit time. A ring pipeline can exhibit four distinct throughput behaviors distinguished by how long it takes a wavefront or a bubble to flow around the ring and by how long it takes all the wavefronts or bubbles in the ring to flow through the cycle in the ring with the slowest period, which will be called the reference cycle. If all cycles in the ring have the same period, then any cycle can be taken as the reference cycle. The throughput of a ring can be wavefront limited, bubble limited, delay limited, or perfectly balanced.

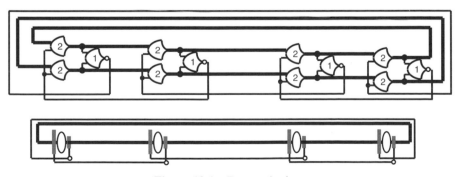

Figure 12.1 Four cycle rings.

Logically Determined Design: Clockless System Design with NULL Convention Logic^{TM}, by Karl M. Fant
ISBN 0-471-68478-3   Copyright © 2005 John Wiley & Sons, Inc.

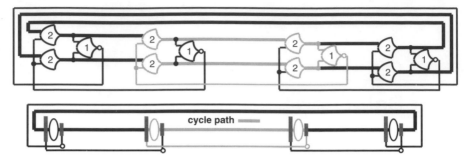

Figure 12.2 Cycle path.

A ring has a population of wavefronts and a population of bubbles. The fundamental question is, When a wavefront and a bubble flow through the reference cycle, can either flow around the ring and rejoin its population before the entire population flows through the reference cycle? This question can be answered in terms of three periods derived from static delay components of the ring: the cycle period, the population period, and the rejoin period.

The cycle period is the sum of the delays around a cycle path as shown in Figure 12.2. For the examples of this chapter the 2 of 2 operator has a delay of 3 tics and the 1 of 2 operator with inverter has a delay of 1 tic. The cycle has a period of 7 tics.

The wavefront population period is the time it takes for the population of wavefronts to flow through the reference cycle, and this is the number of wavefronts in the ring times the cycle period of the reference cycle. The wavefront rejoin period is the time it takes for a wavefront to flow around the ring with no waits, and this is the sum of the delays in the data path of the ring shown in Figure 12.3. The wavefront rejoin period for this 4 cycle ring is $4 \times 3 = 12$ tics.

The bubble population period is the time it takes for the population of bubbles to flow through the reference cycle, and this is the number of bubbles in the ring times

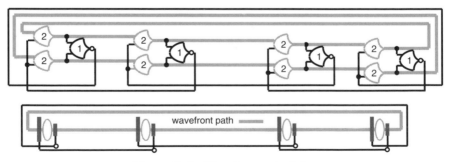

Figure 12.3 Wavefront path of ring.

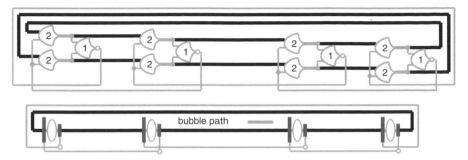

Figure 12.4   Bubble path of ring.

the cycle period of the reference cycle. The bubble rejoin period is the time it takes for a bubble to flow around the ring with no waits, and this is the sum of the bubble path delays in the bubble path of the ring shown in Figure 12.4. The bubble rejoin period for this 4 cycle ring is $4 \times 4 = 16$ tics.

Wavefront limited behavior and bubble limited behavior will be presented in a ring with all identical cycles. If the cycle periods are identical for all cycles in the ring, then the shadow of each cycle is terminated immediately by the shadow of each neighbor. There are no projected shadows, and the behavior of the ring is purely dependent on the structure of the ring and its population of wavefronts and bubbles.

The example ring will be a 24 cycle ring with cycle periods of 7 tics, a wavefront rejoin period of $24 \times 3 = 72$ tics and a bubble rejoin period of $24 \times 4 = 96$ tics.

12.2   WAVEFRONT-LIMITED RING BEHAVIOR

If the wavefront rejoin period is greater than the wavefront population period, then it takes a wavefront longer to flow around the ring than it takes for the population of wavefronts to flow through a reference cycle. The reference cycle will wait on wavefronts, and the throughput of the ring is wavefront limited. If a wavefront can flow around the ring and rejoin its wavefront population before the wavefront population flows through the reference cycle, then there is no throughput limiting wait caused by wavefront flow.

Wavefront limited behavior can be easily understood when there are only two wavefronts in the ring. With two wavefronts in the ring the wavefront population period is $2 \times 7 = 14$ tics, which is much less than the wavefront rejoin period of 72 tics. The two wavefronts flow through the reference cycle, and then the reference cycle must wait for the wavefronts to flow all the way around the ring. The throughput of each cycle waiting on the two wavefronts to flow around the ring and of the ring as a whole is 2 wavefronts every wavefront rejoin period of 72 tics. With four wavefronts in the ring the throughput is four wavefronts every 72 tics.

With two wavefronts in the ring there are 22 bubbles in the ring. The bubble population period is $22 \times 7 = 154$ tics, which is much greater than the bubble rejoin period of 96 tics. Each bubble flows around the ring much faster than all 22 bubbles can flow through any cycle. There is no waiting on bubbles and the ring is not bubble limited.

Table 12.1 profiles the behavior of the example ring with 2 to 24 wavefronts in the ring. The number of bubbles in the ring is 24 minus the number of wavefronts. As wavefronts are added to the ring, the wavefront population period increases and bubble population period decreases while the rejoin periods remain constant. Wavefront-limited behavior continues through 10 wavefronts in the ring. With 12 wavefronts the bubble population period of $12 \times 7 = 84$ tics becomes less than the bubble rejoin period of 96 tics, and the ring becomes bubble limited.

12.2.1   Bubble-limited Ring Behavior

If the bubble rejoin period is greater than the bubble population period, then it takes a bubble longer to flow around the ring than it takes for the population of bubbles to flow through a reference cycle. The reference cycle must wait on bubbles, and the throughput of the ring is bubble limited. If a bubble can flow around the ring and rejoin its bubble population before the bubble population flows through the reference cycle, then there is no throughput limiting wait caused by bubble flow.

Bubble-limited behavior can be easily understood when there are 22 wavefronts in the ring with only two bubbles. With two bubbles in the ring the bubble population period is $2 \times 7 = 14$ tics, which is much less than the bubble rejoin period of 96 tics. The two bubbles flow through any cycle much faster than they flow around the ring. The two bubbles flow through the reference cycle, and then the reference cycle must wait for the bubbles to flow all the way around the ring. As each bubble passes a point in the ring, a wavefront flows through the bubbles and advances one cycle in the ring. Looking at any point in the ring, one will observe the flow of two wavefronts each time the bubbles pass. So the throughput of each cycle waiting on the two bubbles to flow around the ring, and of the ring as a whole, is 2 wavefronts every bubble rejoin period of 96 tics. With four bubbles in the ring the throughput is four wavefronts every 96 tics.

Notice that the peak throughput occurs with a cycles per wavefront ratio of 2.4. To sustain continuous flow, each wavefront must always have a bubble to flow into. This means that the peak throughput occurs when the ring is approximately half full of wavefronts and half full of bubbles at around a cycles per wavefront ratio of 2.00. The peak at 2.40 in this example is due to the imbalance of the rejoin periods.

12.2.2   Delay-limited Ring Behavior

If there is a cycle with a slower period than the other cycles, the ring can enter a mode where both wavefronts and bubbles flow around the ring faster than their population can flow through the slow cycle. Then the ring is neither wavefront

TABLE 12.1 Behavior Profile for Example Ring with Varying Wavefront Populations

Wavefronts in Ring	Wavefront Population Period	Wavefront Rejoin Period	Bubbles in Ring	Bubble Population Period	Bubble Rejoin Period	Cycles/ Wavefront	Throughput Waves/Period	Throughput Waves/100 tics	Limiting Behavior Mode
2	14	72	22	154	96	12.00	2/72	2.78	Wavefront
4	28	72	20	140	96	6.00	4/72	5.56	Wavefront
6	42	72	18	126	96	4.00	6/72	8.33	Wavefront
8	56	72	16	112	96	3.00	8/72	11.11	Wavefront
10	70	72	14	98	96	2.40	10/72	13.89	Wavefront
12	84	72	12	84	96	2.00	12/96	12.50	Bubble
14	98	72	10	70	96	1.71	10/96	10.42	Bubble
16	112	72	8	56	96	1.50	8/96	8.33	Bubble
18	126	72	6	42	96	1.33	6/96	6.25	Bubble
20	140	72	4	28	96	1.20	4/96	4.16	Bubble
22	154	72	2	14	96	1.09	2/96	2.08	Bubble
24	168	72	0	0	96	1.00	Deadlock	Deadlock	Deadlock

limited nor bubble limited. The throughput limiting wait is caused by the delay of the cycle with the slowest period and the throughput of the ring is delay limited.

Consider a 4 tic delay added to the data path of one cycle, giving it a period of 11 tics. All the other cycles in the ring retain periods of 7 tics. This 4 tic delay in the data path will also add 4 tics to the wavefront rejoin period, which becomes 76 tics. The bubble rejoin period, which does not include the data path, remains 96 tics.

Table 12.2 show the profile of the ring behavior with the slow 11 tic cycle. Notice that population periods increase faster with the larger cycle period and more quickly overtake their respective rejoin periods. Consequently there are configurations where the ring is neither wavefront limited nor bubble limited but is limited by the throughput of the slow cycle. The throughput of a ring can be no greater than the throughput of its slowest cycle. For 8, 10, 12, and 14 wavefronts in the ring, the ring is delay limited and the throughput is constant.

In delay-limited mode a remarkable behavior occurs. The wavefront shadows and the bubble shadows of the slow cycle project around the ring unchallenged, meet on the opposite side of the ring, and exactly cancel. The waits imposed by the shadows cause every cycle in the ring to cycle with the same period as the slow cycle. These shadows cause all the wavefronts in the ring to have a uniform period and to be uniformly distributed independent of the wavefront population. As the wavefront population varies within the delay-limited behavior, no change whatever occurs in the behavior of the ring. This behavior is shown later in Figures 12.9 through 12.12.

One just knows that the wavefronts and the bubbles are racing around the ring and jamming up around the slow cycle, and that the jam varies as the population of wavefronts varies. One would not possibly guess from this mistaken intuitive picture that the signals of the cycles could look uniformly distributed and constant for all cases. But the shadows enforce a wait discipline on the ring, and the wavefronts do indeed flow around the ring in perfect coordination and distribution. This delay-limited behavior of the ring is utterly counter intuitive.

When it takes longer for a wavefront to travel around the ring than it does for the population of wavefronts to get through the slow cycle, then the slow cycle has to wait longer than its cycle period. Its shadow is overcome by this wait and the ring enters wavefront limited behavior. Similarly, when a bubble takes longer to travel around the ring than the population of bubbles takes to flow through the slow cycle, then the ring enters bubble limited behavior. The specific behavior causing the longest wait dominates the behavior of the ring.

12.2.3 Perfectly Balanced Ring Behavior

When the population periods and the rejoin periods are equal and the cycle periods are all equal, the behavior of the ring is perfectly balanced. This behavior is only of academic interest in that it cannot be expected to be achieved in a real pipeline ring. Consider a 1 tic delay added to the data path of each cycle in the ring. Each cycle then contributes 4 tics to the data path and 4 tics of delay to the bubble path.

TABLE 12.2 Behavior Profile for Example Ring with 4 tic Delay in One Cycle

Wavefronts in Ring	Wavefront Population Period	Wavefront Rejoin Period	Bubbles in Ring	Bubble Population Period	Bubble Rejoin Period	Cycles/ Wavefront	Throughput Waves/Period	Throughput Waves/100 tics	Limiting Behavior Mode
2	22	76	22	242	96	12.00	2/76	2.63	Wavefront
4	44	76	20	220	96	6.00	4/76	5.26	Wavefront
6	66	76	18	198	96	4.00	6/76	7.89	Wavefront
8	88	76	16	176	96	3.00	1/11	9.09	Delay
10	110	76	14	154	96	2.40	1/11	9.09	Delay
12	132	76	12	132	96	2.00	1/11	9.09	Delay
14	154	76	10	110	96	1.71	1/11	9.09	Delay
16	176	76	8	88	96	1.50	8/96	8.33	Bubble
18	198	76	6	66	96	1.33	6/96	6.25	Bubble
20	220	76	4	44	96	1.20	4/96	4.16	Bubble
22	242	76	2	22	96	1.09	2/96	2.08	Bubble
24	264	76	0	0	96	1.00	Deadlock	Deadlock	Deadlock

TABLE 12.3 Behavior Profile for Example Ring with Balanced Rejoin Periods

Wave fronts in Ring	Wavefront Population Period	Wavefront Rejoin Period	Bubbles in Ring	Bubble Population Period	Bubbles Rejoin Period	Cycles/ Wavefront	Throughput Waves/Period	Throughput Waves/100 tics	Limiting Behavior Mode
2	16	96	22	176	96	12.00	2/96	2.08	Wavefront
4	32	96	20	160	96	6.00	4/96	4.16	Wavefront
6	48	96	18	144	96	4.00	6/96	6.25	Wavefront
8	64	96	16	128	96	3.00	8/96	8.33	Wavefront
10	80	96	14	112	96	2.40	10/96	10.42	Wavefront
12	96	96	12	96	96	2.00	12/96	12.5	Balanced
14	112	96	10	80	96	1.71	10/96	10.42	Bubble
16	128	96	8	64	96	1.50	8/96	8.33	Bubble
18	144	96	6	48	96	1.33	6/96	6.25	Bubble
20	160	96	4	32	96	1.20	4/96	4.16	Bubble
22	176	96	2	16	96	1.09	2/96	2.08	Bubble
24	192	96	0	0	96	1.00	Deadlock	Deadlock	Deadlock

Both rejoin periods will then be 96 tics. The behavior profile of the ring is presented in Table 12.3.

Notice that the throughput variation is symmetric, that the peak occurs at exactly 2.0 cycles per wavefront, and that at this point the wavefront flow and the bubble flow are perfectly balanced and flowing around the ring with no waits. All signals coordinate just in time for perfectly efficient behavior.

12.3 THE CYCLE-TO-WAVEFRONT RATIO

It is convenient to view ring behavior in terms of the ratio of the number of cycles in the ring to the number of wavefronts in the ring. Figure 12.5 illustrates this ratio of throughout performance. Peak performance occurs around two cycles per wavefront. Bubble-limited behavior occurs when there are less than two cycles per wavefront. Wavefront-limited behavior occurs when there are more than two cycles per wavefront. Performance falls off precipitously to deadlock with bubble-limited behavior, so it is a good idea to steer clear of bubble-limited behavior in designing rings. Throughput falls off much more gradually with wavefront-limited behavior and is a much more forgiving region of behavior. If one keeps adding cycles to a ring with a constant number of wavefronts, the throughput of the ring asymptotically approaches zero.

Delay-limited behavior imposes a throughput limiting plateau on the performance curve. One can operate anywhere on the plateau with identical performance.

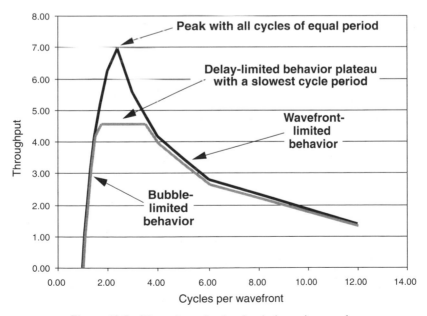

Figure 12.5 Throughput for the ring in its various modes.

As there are usually slow function cycles in any ring structure, the plateau provides an easily identifiable region for conveniently flexible design. Optimal configuration of a ring occurs on the plateau closest to the bubble-limited cliff, which provides maximal throughput with minimal cycles. However, if there is a large variation of behavior, one should be careful not to fall off the cliff into bubble-limited behavior.

With a given wavefront population for a ring the performance of the ring can be tuned by adding or removing buffer cycles. A ring is most efficient when functional cycles can be partitioned into smaller cycles without increasing the wavefront population.

If there are variable delay behavior cycles in the ring, then the graph will become animated. The delay-limited plateau will bounce up and down, and with a small plateau the ring could even temporarily transition between bubble, and wavefront-limited modes.

12.4 RING SIGNAL BEHAVIOR

It is important to understand the signal behavior associated with each ring behavior. Figures 12.6 through 12.17 illustrate the view of pipeline ring behavior in terms of the signal behavior of the individual cycles. Each figure represents a particular population of wavefronts and a particular behavior mode. Each figure shows the signal trace for each cycle of a 24 cycle ring. The signals are captured after the start-up

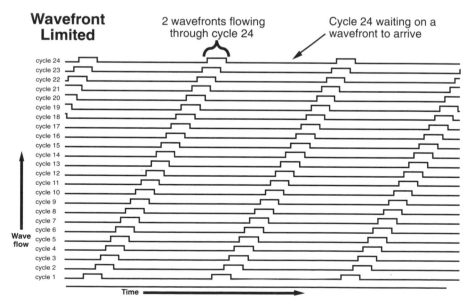

Figure 12.6 Two wavefronts and 22 bubbles flowing through a 24 cycle ring.

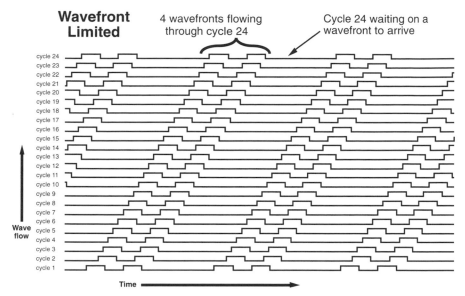

Figure 12.7 Four wavefronts and 20 bubbles flowing through a 24 cycle ring.

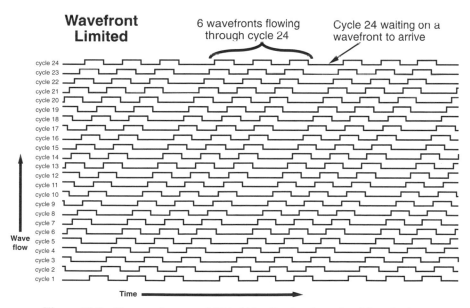

Figure 12.8 Six wavefronts and 18 bubbles flowing through a 24 cycle ring.

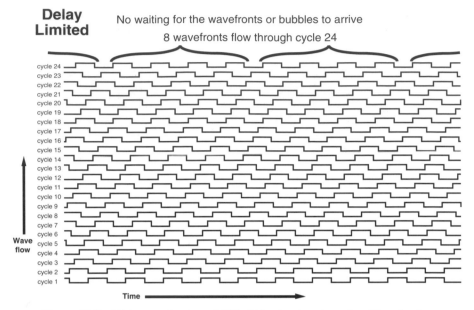

Figure 12.9   Eight wavefronts and 16 bubbles flowing through a 24 cycle ring.

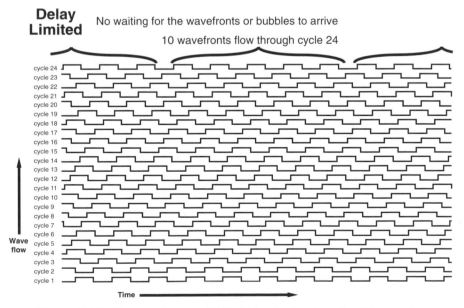

Figure 12.10   Ten wavefronts and 14 bubbles flowing through a 24 cycle ring.

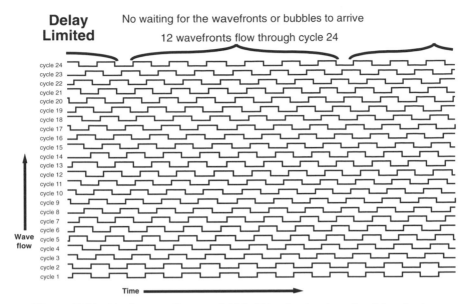

Figure 12.11   Twelve wavefronts and 12 bubbles flowing through a 24 cycle ring.

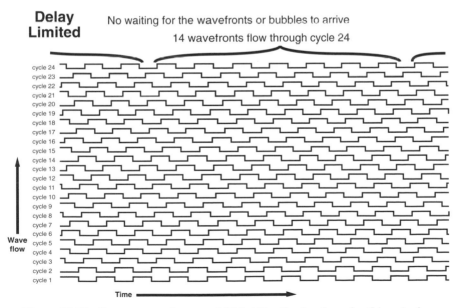

Figure 12.12   Fourteen wavefronts and 10 bubbles flowing through a 24 cycle ring.

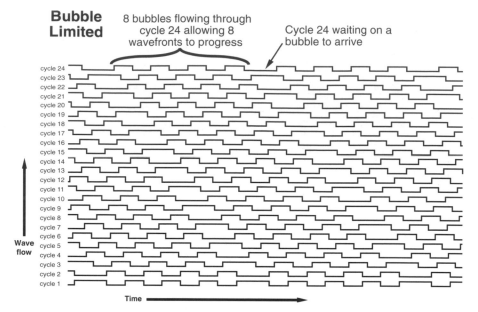

Figure 12.13   Sixteen wavefronts and 8 bubbles flowing through a 24 cycle ring.

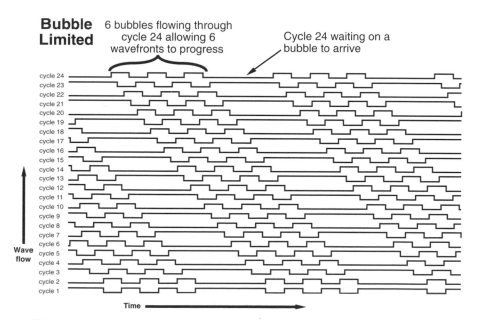

Figure 12.14   Eighteen wavefronts and 6 bubbles flowing through a 24 cycle ring.

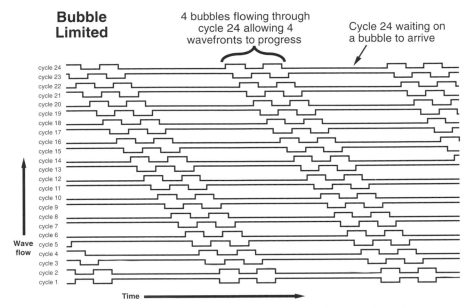

Figure 12.15   Twenty wavefronts and 4 bubbles flowing through a 24 cycle ring.

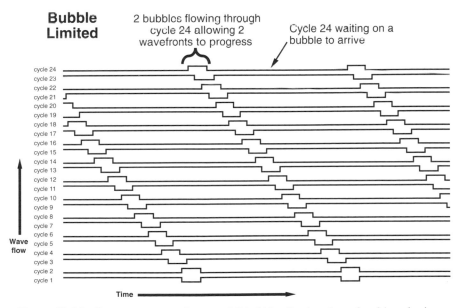

Figure 12.16   Twenty-two wavefronts and 2 bubbles flowing through a 24 cycle ring.

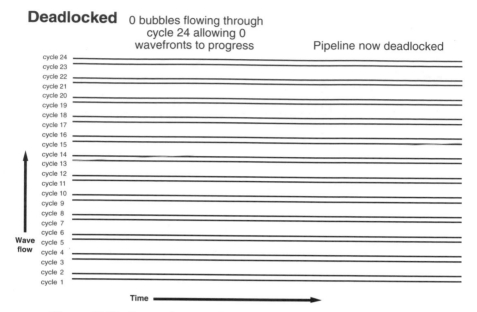

Figure 12.17 Twenty-four wavefronts not flowing through a 24 cycle ring.

transient when the ring behavior has stabilized. Notice that in delay-limited behavior with 8 to 14 wavefronts the signal behavior of the cycles and the throughput of the ring are identical and constant. Notice also that in wavefront-limited behavior every cycle waits in the same state, whereas in bubble-limited behavior alternate cycles wait in alternating states.

CHAPTER 13

Interacting Pipeline Structures

A logically determined system is a structure of pipelines. Once the logical behavior is determined, the next question is the performance of the system in terms of through-put. How do pipelines interact to affect throughput, and how can pipeline structures be designed to achieve optimal throughput? Given the dynamic behavior of the cycles and pipelines, one might think that an intuitive understanding of the behavior of and design of a multi-pipeline structure would be impossible. The only approach to the reliable design of multi-pipeline structures would seem to be through complex mathematical network and queueing theories supplemented with detailed simulation studies.

It is possible, however, to characterize the behavior of a multi-pipeline structure solely in terms of static relationships. These static relationships are easily under-standable and are sufficient to support precise synthesis of pipeline structures with optimal performance. The principles of the static synthesis of multiple-pipeline structures will be presented with several examples. This chapter will cover the basics of two-pipeline structures. The next chapter will show how to construct complex pipeline structures in terms of two-pipeline structures.

A two-pipeline structure consists of two pipelines combined by a fan-out structure and by a fan-in structure through a combinational expression as shown in Figure 13.1.

There are two points of synchronization. Where the pipelines fan-out, they are synchronized through the acknowledge paths (bubbles) of the two fan-out cycles.

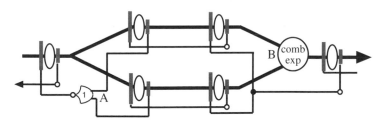

Figure 13.1 Base model of two-pipeline structure.

Logically Determined Design: Clockless System Design with NULL Convention Logic^{TM}, by Karl M. Fant
ISBN 0-471-68478-3 Copyright © 2005 John Wiley & Sons, Inc.

This is shown as operator A in Figure 13.1. Where the pipelines fan-in, they are synchronized through the data path by a combinational expression B in Figure 13.1. The combined pipelines form a closed structure bounded by the points of synchronization. The behavior of the closed structure is limited by the structure itself, similarly to the way that the behavior of a ring is limited by its closed structure.

13.1 PRELIMINARIES

For clarity of presentation, collapsed data path models will be used for the examples. The behaviors exhibited and the synthesis techniques discussed are fully applicable to pipeline structures with full-width data paths.

The input of each two pipeline structure is simulated with an auto-produce cycle and the output is simulated with an auto-consume cycle. The terminology is in relation to wavefronts. The auto-produce cycle will produce a wavefront and consume a bubble every time a bubble flows into the cycle. The auto-consume cycle will consume a wavefront and produce a bubble every time a wavefront flows into the cycle. The auto cycles shown in Figure 13.2 will always cycle faster than any other cycle in a pipeline structure, will not cast shadows into the structure, and so will not affect the inherent throughput behavior of the structure. The auto cycles produce a steady stream of wavefronts and bubbles on demand allowing the isolated analysis of the inherent throughput behavior of a pipeline structure.

The unit of propagation delay for the examples is a tic. Each operator has a propagation delay of 3 tics for both DATA and NULL. An inverter has a propagation delay of 1 tic for both DATA and NULL. Specific delays are shown as circles with the value in tics of the delay in the circle. Wires have no delay. What will be important is propagation delays along specific paths. It is simple to include wire delays on these paths for real world structures, but for purposes of illustration it is simpler to leave them out. The characterizations of these delays do not always need to be either precise or absolute. In many cases there will be a range of structural configurations that will deliver an optimal throughput for a range of delay parameters. The delays can also be conveniently relative. For instance, the delays in tics mentioned above can be interpreted to mean that the delay of an operator is three times longer than the delay of an inverter.

Figure 13.2 Auto cycles for simulating isolated pipeline structures.

13.2 EXAMPLE 1: THE BASICS OF A TWO-PIPELINE STRUCTURE

The initial discussion will focus on a simple example of a concurrent-path function. This might be a function something like $F = (2y - 1)^2 + 3y$. Each term is computed in a concurrent path branching from the y input and then combined through a sum to the final result. The computations will be characterized abstractly in a collapsed data path model with delay elements representing function components. The final add is represented by a 2 of 2 operator. The example will assume a lower data path with one computation component and an upper data path with three computation components. The preliminary structure is shown in Figure 13.3 as a pipeline structure with three cycles; a single computation cycle encompassing the large combinational expression, an auto-produce cycle, and an auto-consume cycle. The throughput of the structure is determined by the cycle period of the computation cycle. This cycle period is determined by the upper data path, which takes longer than the lower data path. The cycle period is 22 tics and the throughput of the structure is one wavefront every 22 tics.

A pipeline structure is created by pipelining the upper and lower data paths with a cycle for each function delay. This results in a pipeline structure with an upper pipeline of five cycles and a lower pipeline of three cycles as shown in Figure 13.4. The fan-out bubble synchronization is operator A, and the fan-in wavefront synchronization is operator B.

The pipelining has not changed the functionality of the expression but what now is the throughput of the structure? Can the throughput be enhanced by adding buffer cycles to one of the pipelines. If so which pipeline, how many cycles, and where in the pipeline should they be placed?

13.2.1 Basics of Flow

Each population of wavefronts and bubbles in the structure flows out of the structure, and each population is renewed within the structure by wavefronts flowing in

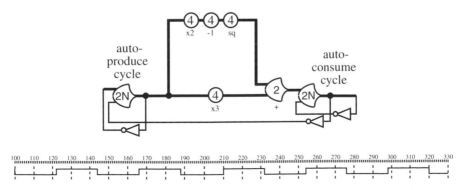

Figure 13.3 Example function mapped into two data paths in a single cycle.

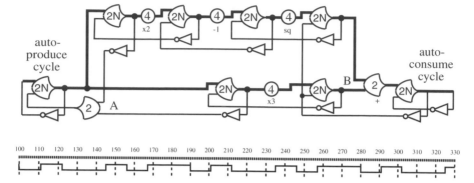

Figure 13.4   Pipeline structure derived from combinational function.

through the input and by bubbles flowing in through the output. Bubbles and wave-fronts flow out of the structure through the synchronization points. When there is a bubble from each pipeline presented to operator A, a bubble will flow out of the structure and a wavefront will flow into the structure and be copied to both pipelines. When there is a wavefront from each pipeline presented to operator B, a wavefront will flow out of the structure and a bubble will flow into the structure and be copied to both pipelines. The quantity of bubbles and wavefronts in each pipeline will always be a constant equal to the number of cycles in the pipeline.

The cycles of each pipeline of the example are initialized to bubbles. There are no wavefronts initialized in the structure. After initialization, bubbles from each pipe-line will synchronize and begin flowing out of the structure, allowing wavefronts to flow into the structure. This flow will continue until the smallest population of bubbles is depleted. In this example after the three bubbles of the lower pipeline have synchronized with three bubbles of the upper pipeline and allowed three wave-fronts into the structure, the lower pipeline has run out of it its initial supply of bubbles. The fourth bubble of the lower pipeline that will synchronize with the fourth bubble of the upper pipeline must be a new bubble that entered the structure when a wavefront flowed out of the structure. At this point there may or may not be a wait for the fourth bubble in the lower pipeline. The arrival of this fourth bubble depends on the exit of the first bubble and on the structure itself.

As the first bubble flows out the input of the pipeline, it allows a first wavefront into the pipeline that will propagate to the output of the structure and allow a fourth bubble to flow into the structure. This bubble will be copied to the lower pipeline as the fourth bubble of the lower pipeline and will be copied to the upper pipeline as the sixth bubble of the upper pipeline. This fourth/sixth bubble, which is the first bubble to enter the structure, is the renewal bubble for the first bubble that left the structure. This fourth bubble in the lower pipeline will then flow to the input of the structure. The critical question is whether this fourth bubble arrives at the input of the lower pipeline in time to flow out as the fourth bubble of the structure with no wait at the synchronization operator A or whether it arrives after the three bubbles have flowed out and the fourth bubble of the upper pipeline has to wait for the fourth bubble of the lower pipeline. This fourth bubble in the lower pipeline will be called the renewal bubble for the first bubble of the lower pipeline.

This fourth bubble of the lower pipeline that was allowed into the structure by the first exiting wavefront will synchronize, flow out, and allow in a fourth wavefront, which is the renewal wavefront for the first wavefront that exited the structure. The critical question is whether this fourth wavefront arrives at the output in time to flow out as the fourth wavefront of the structure with no wait at the synchronization operator B or whether it arrives after the three wavefronts have flowed out and the synchronization operator B has to wait for the fourth wavefront.

The answer to these two critical questions can be formulated in terms of two periods: the population period, which is the time it takes for a population to flow out of the structure, and the renewal period, which is the time it takes from when a population member flows out of the structure to the time when its renewal member can flow out of the structure. There are four populations in the structure. Each pipeline has a bubble population and a wavefront population.

Renewal-limited Behavior. If a population period is smaller than its renewal period, then the population flows out of the structure before the renewal member can catch up, a wait occurs, and the throughput of the structure is reduced. This will be called renewal-limited behavior.

Delay-limited Behavior. If all population periods are greater than or equal to their renewal periods, then all the renewal members catch up with the populations before they can flow out of the structure, no renewal wait occurs, and the structure flows at its maximum throughput limited only by the delays of the cycles in the pipeline. This will be called delay-limited behavior.

The Maximum Throughput. Absent any waits on renewal members, wavefronts and bubbles will flow through the structure at its maximum throughput. The maximum throughput of a structure (the throughput plateau) is the throughput of its slowest cycle. Just as with any pipeline structure, a slowest cycle will project a shadow through the structure, limiting its throughput to the throughput of that cycle. The slowest cycles of the example structure are the cycles with the 4 tic delay, which have cycle periods of 11 tics. So the maximum throughput of the structure is one wavefront every 11 tics.

The Wavefront Population. The wavefront population of each pipeline is the number of wavefronts that can occupy the pipeline. This is determined by the wavefronts initialized in the pipeline and by the smallest population of bubbles in the structure. In the current example there are no wavefronts initialized, and the smallest population of bubbles is the three bubbles of the lower pipeline. The three bubbles will flow out of the structure, allowing three wavefronts into each pipeline. After that, a new bubble must flow into the structure to allow another wavefront into the structure, but a wavefront must flow out of the structure to allow this new bubble into the structure. Consequently, there will never be more than three wavefronts in either pipeline.

If wavefronts were initialized in a pipeline, then the wavefront population would be the initialized wavefronts plus the wavefronts allowed by the smallest population of bubbles.

The Bubble Population. The bubble population is the number of bubbles that can occupy a pipeline. This is determined by the bubbles initialized in the pipeline. A bubble population could be affected by the smallest population of initialized wavefronts. But because it rarely makes sense to initialize wavefronts in both pipelines, the smallest population of wavefronts is always zero.

The Excess Population. The two extra bubbles of the upper pipeline will be called an excess population. There will never be less than two bubbles in the upper pipeline. If wavefronts are initialized in one pipeline, then they are an excess population.

The Population Period. A population will flow at the maximum throughput of the structure. A population period is the number of members of the population times the period of the slowest cycle. For the current example the wavefront population for both pipelines is three with wavefront population period of 33 tics. The bubble population of the upper pipeline is five with a population period of 55 tics, and the bubble population of the lower pipeline is three with a population period of 33 tics.

The Renewal Period. The renewal period is the time it takes from when a member of a population leaves the structure to when its renewal member is itself ready to exit the structure. The renewal period is determined by the renewal path.

The Renewal Path. The first wavefront allowed in by the first bubble flows through bubbles to the output and encounters no waits. The renewal bubble allowed in by the this wavefront flows through wavefronts until it catches up with its population of bubbles, or until it reaches the input. If a renewal bubble does not catch up with its population, it never encounters a wait along its flow path. In trying to catch up with their populations, bubbles flow without hindrance through wavefronts and wavefronts flow without hindrance through bubbles. This means that the renewal period can be determined solely in terms of the delays along the renewal path.

The renewal path is the data path over which a wavefront flows from the input to the output and the acknowledge path over which a bubble flows from the output to the input. There are four possible renewal paths in a two-pipeline structure constructed from combinations of two possible wavefront paths (data paths) and two possible bubble paths (acknowledge paths). The possible renewal paths for the example of Figure 13.4 are shown in Figure 13.5. The renewal path symbol indicates the structure of the renewal path. The wide line is the wavefront path and the narrow line is the bubble path.

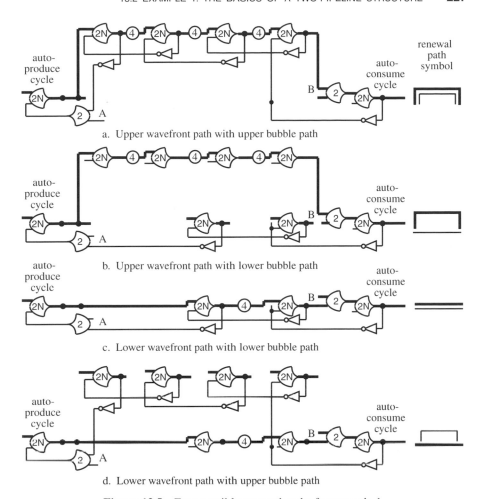

Figure 13.5 Four possible renewal paths for example 1.

In the renewal path of Figure 13.5*a*, the wavefront path is the data path of the upper pipeline, and the bubble path is the acknowledge path of the upper pipeline. The four completeness operators in the upper pipeline must be counted twice, once for the wavefront propagation and once for the bubble propagation. The renewal period for this path is 53 tics.

In the renewal path of Figure 13.5*b*, the wavefront path is the data path of the upper pipeline, and the bubble path is the acknowledge path of the lower pipeline. The renewal period for this path is 45 tics.

In the renewal path of Figure 13.5*c*, the wavefront path is the data path of the lower pipeline, and the bubble path is the acknowledge path of the lower pipeline. The two completeness operators in the lower pipeline must be counted twice, once for the wavefront propagation and once for the bubble propagation. The renewal period for this path is 31 tics.

In the renewal path of Figure 13.5*d*, the wavefront path is the data path of the lower pipeline, and the bubble path is the acknowledge path of the upper pipeline. The renewal period for this path is 39 tics.

Determining the Renewal Path. Each population of wavefronts and bubbles in the structure must be renewed and has a renewal path. Each renewal path has a wavefront segment and a bubble segment. The second segment of the renewal path must be the path of the population itself. The first segment of the renewal path is determined by propagation to synchronization behavior.

Consider the wavefront renewal paths for the example. The first wavefront flows out of the structure, allowing a bubble to enter that is the fourth bubble of the lower pipeline and the sixth bubble for the upper pipeline. The critical question about this bubble is when will it allow a renewal wavefront into the structure. This will occur when the bubble propagates through the lower pipeline and synchronizes with the fourth bubble of the upper pipeline, allowing the renewal wavefront into the structure that is copied to each pipeline data path. The copy of the bubble in the upper pipeline simply joins the bubble population and does not participate in allowing the renewal wavefront into the structure. So the first segment bubble portion of the renewal path for both wavefront populations is the acknowledge path of the lower pipeline. The second segment is the data path of each wavefront population.

So the renewal path for the wavefront population of the upper pipeline is the acknowledge path of the lower pipeline and the data path of the upper pipeline that is the configuration of Figure 13.5*b*. The renewal period of this path is 45 tics. The wavefront population period is 33 tics, so the wavefront population of the upper pipeline does not get renewed in time and imposes a wait on the structure affecting its throughput.

The renewal path for the wavefront population of the lower pipeline is the acknowledge path of the lower pipeline and the data path of the lower pipeline. This is the configuration of Figure 13.5*c*, with a renewal period of 32 tics. So the wavefront population of the lower pipeline is not renewal limited.

Consider the bubble renewal paths for the example. The first bubble flows out of the structure, allowing a wavefront to enter that is the first wavefront of the structure. The critical question about this wavefront is when will it allow a renewal bubble into the structure.

There is no excess population of wavefronts in the current example. When a wavefront enters the input of the structure and is copied to both pipelines, the copies must flow through their respective data paths and synchronize with each other at the output. In this case the path that determines the time from entry to synchronization and exit is the data path with the longest delay, which in this case is the data path of the upper pipeline. So the first segment of the renewal path for both bubble populations is the data path of the upper pipeline.

The renewal path for the bubble population of the upper pipeline is the data path of the upper pipeline and the acknowledge path of the upper pipeline. This is the configuration of Figure 13.5*a*, which has a renewal period of 53 tics. The bubble population period of the upper pipeline is 55 tics, so the population is not renewal limited.

The renewal path for the bubble population of the lower pipeline is the data path of the upper pipeline and the acknowledge path of the lower pipeline. This is the configuration of Figure 13.5*b*, which has a renewal period of 45 tics. The bubble population period of the lower pipeline is 33 tics, so the population is renewal limited. The renewal paths of the structure with three cycles in the lower pipeline are summarized in Figure 13.6.

Any population that is not renewed in time will cause a wait that affects the throughput of the structure. The wavefronts of the upper pipeline and the bubbles of the lower pipeline are not renewed in time. Three wavefronts and bubbles flow in 33 tics, and then the structure must wait another 12 tics before another three wavefronts and bubbles will flow. The throughput of the structure is three wavefronts every 45 tics, as shown by the simulation signal trace of Figure 13.4.

Can the structure be configured to achieve its maximum throughput of one wavefront every 11 tics?

13.2.2 Increasing the Throughput

The goal is to make all population periods greater than or equal to their renewal periods. There are two options. Renewal periods can be reduced, or population periods can be increased. It may be difficult to impossible to configure delays to change renewal periods for a given cycle structure, but it is easy to configure a cycle structure to change population periods with a given delay constraint. Adding buffer cycles increases populations and hence their periods. The question is where to add the buffer cycles in the structure.

The population of bubbles in the lower pipeline and the population of wavefronts in the upper pipeline are renewal limited. The population of wavefronts in the upper pipeline is dependent on the population of bubbles in the lower pipeline. Increasing the population of bubbles in the lower pipeline will increase the population of wavefronts in the upper pipeline and hence both population periods.

Adding a buffer cycle to the lower pipeline increases the smallest population of bubbles, increases the population of wavefronts, and increases both population periods. But adding a cycle also increases the renewal period for each population. While the population period increases by 11 tics, the renewal period increases by

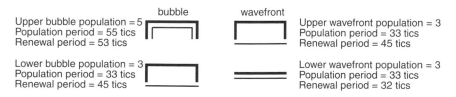

Upper bubble population = 5
Population period = 55 tics
Renewal period = 53 tics

bubble

Upper wavefront population = 3
Population period = 33 tics
Renewal period = 45 tics

wavefront

Lower bubble population = 3
Population period = 33 tics
Renewal period = 45 tics

Lower wavefront population = 3
Population period = 33 tics
Renewal period = 32 tics

Figure 13.6 Renewal paths with three cycles in lower pipeline.

only 4 tics. With each cycle, the population period gains an advantage of 7 tics over the renewal period. With a fourth cycle added to the lower pipeline, both population periods become 44 tics, and the renewal period for each becomes 49 tics. With a fifth cycle added as shown in Figure 13.7, the population periods become 55 tics and the renewal periods become 53 tics. The population period becomes greater than the renewal period. All the renewal wavefronts and bubbles catch up with their populations before they flow out of the structure, enabling the populations of the structure to flow continually, with no waits, at the maximum throughput of the structure.

How did adding cycles to the lower pipeline affect the other populations? Since the data path of the upper pipeline remains the data path with the longest delay, the renewal path for the bubble population of the upper pipeline remains the same. Both path segments are in the upper pipeline, and adding a cycle to the lower pipeline had no effect at all on the bubble population of the upper pipeline. Its population period remains 55 tics and its renewal period remains 53 tics.

The renewal period for the wavefront population of the lower pipeline was 32 tics, and its population period was 33 tics. Its renewal path was entirely in the lower pipeline, so adding cycles to the lower pipeline will affect this renewal path. But the added cycles affected the wavefront population. While each additional cycle added 7 tics to the renewal period it added 11 tics to the population period. So the population period stayed greater than the renewal period.

With five cycles in the lower pipeline all population periods in the structure, summarized in Figure 13.8, are greater than their renewal periods, and the structure becomes delay limited with a throughput of one wavefront every 11 tics. It does not matter where in the lower pipeline the buffer cycles are placed. The important factor is that there are initially five bubbles in the lower pipeline.

As more cycles are added to the lower pipeline, the bubble population of the lower pipeline continues to increase, and its population period stays greater than its renewal period. Because the five bubbles of the upper population are now the smallest bubble population, they determine the wavefront population which will remain at five wavefronts. Since the wavefront population will remain at five

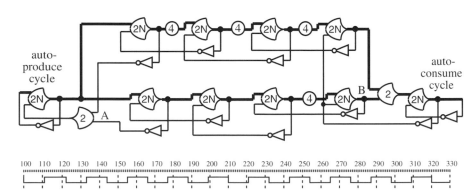

Figure 13.7   Five cycles in lower pipeline.

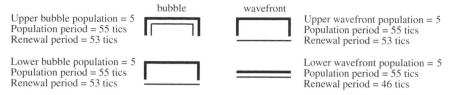

Figure 13.8   Renewal paths with five cycles in the lower pipeline.

wavefronts while the renewal paths increase, eventually the renewal period will exceed the wavefront population period of the lower pipeline, and the structure will become renewal limited.

With a sixth cycle added to the lower pipeline, the lower pipeline now has an excess bubble population. The bubble portion of the renewal path for the wavefront populations becomes the acknowledge path of the upper pipeline. The new renewal path configuration is shown in Figure 13.9. All population periods remain greater than their renewal periods, and the structure remains delay limited.

With eight cycles in the lower pipeline, shown in Figure 13.10, another change occurs in the renewal paths. The delay of the data path of the lower pipeline, 25 tics, becomes longer than the delay of the upper pipeline data path, 24 tics, and the wavefront portion of the renewal path of the bubble populations becomes the lower pipeline data path, as shown in Figure 13.11. Although the renewal path

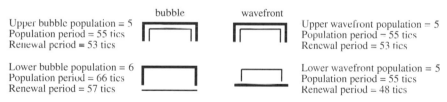

Figure 13.9   Renewal paths with six cycles in the lower pipeline.

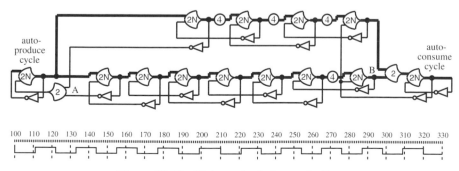

Figure 13.10   Eight cycles in lower pipeline.

bubble   wavefront

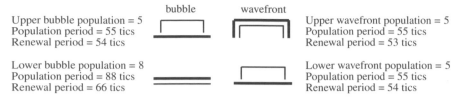

Upper bubble population = 5
Population period = 55 tics
Renewal period = 54 tics

Upper wavefront population = 5
Population period = 55 tics
Renewal period = 53 tics

Lower bubble population = 8
Population period = 88 tics
Renewal period = 66 tics

Lower wavefront population = 5
Population period = 55 tics
Renewal period = 54 tics

Figure 13.11   Renewal paths with eight cycles in the lower pipeline.

has changed, the new renewal period of 54 tics is still smaller than the population period of 55 tics, and the behavior of the structure remains delay limited with a throughput of one wavefront every 11 tics.

When the ninth cycle is added to the lower pipeline in Figure 13.12, the renewal period for the bubble population of the upper pipeline and the wavefront population of the lower pipeline becomes 57 tics and exceeds the population periods of 55 tics. The behavior becomes renewal limited and the throughput becomes five wavefronts every 57 tics. From this point as more cycles are added to the lower pipeline, the renewal period increases and the throughput decreases.

Table 13.1 profiles the behaviors of the configurations for example 1. The table includes configurations with one and two cycles in the lower pipeline. Figure 13.13 shows the waveforms of each configuration, and Figure 13.14 charts the throughput for the configurations.

13.2.3   Summary of Example 1

Pipelining a structure does not in and of itself deliver throughput performance. The structure must be properly configured. The behavior of the structure can be characterized in terms of the period of the slowest cycle, the population periods, and the renewal periods. All periods are derived from the structure itself and the static delay components of the structure.

The optimal configuration of a two-pipeline structure can be discovered by constructing a behavior table of these periods and parsing the table to make design

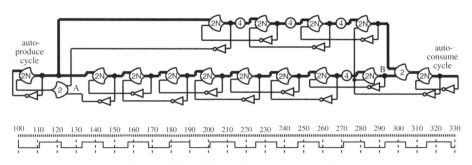

Figure 13.12   Nine cycles in lower pipeline.

TABLE 13.1   Example 1 Behavior Profile

Cycles in Lower Pipeline	Limiting Cycle Period	Relevant Population Period	Relevant Renewal Period	Throughput Wave/Period	Throughput Wave/ 100 tics	Limiting Behavior Mode
No pipes	22	—	—	1/22	4.545	—
1	17	17	37	1/37	2.703	Renewal
2	14	28	41	2/41	4.878	Renewal
3	11	33	45	3/45	6.667	Renewal
4	11	44	49	4/49	8.163	Renewal
5	11			1/11	9.090	Delay
6	11			1/11	9.090	Delay
7	11			1/11	9.090	Delay
8	11			1/11	9.090	Delay
9	11	55	57	5/57	8.772	Renewal
10	11	55	60	5/60	8.333	Renewal
11	11	55	63	5/63	7.937	Renewal

decisions involving simple rules. For this example five cycles in each pipeline formed the most efficient structure, providing the optimal throughput. It is easy to overallocate cycles. The right edge of the plateau delivers the same throughput but uses unnecessary cycles and switching.

This example was extended to its extremes of behavior to elucidate the range of behavior. Normally one would terminate the table when the known optimal throughput is achieved with minimal cycles.

13.3   EXAMPLE 2: A WAVEFRONT DELAY STRUCTURE

Example 2 is a wavefront delay structure. A delay relationship is created by initializing wavefronts in one pipeline of a two pipeline structure. Consider a structure

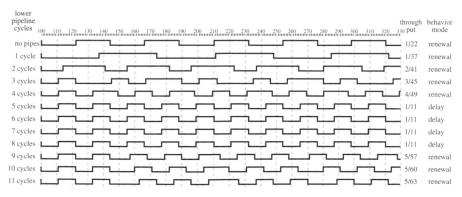

Figure 13.13   Data path waveforms for the configurations of Example 1.

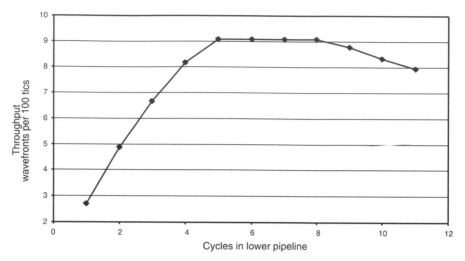

Figure 13.14   Throughput behavior for the configurations for Example 1.

with two wavefronts (DATA/NULL pair of wavefronts) initialized in the upper pipeline. When a wavefront enters the structure and is copied to both pipelines, the upper copy of the wavefront joins the population of wavefronts, and the lower wavefront synchronizes with and interacts with the next wavefront of the population which is the wavefront that entered the structure two wavefronts prior to the wavefront in the lower pipeline. The upper pipeline provides a two-wavefront delay (one DATA–NULL pair) in relation to the lower pipeline. Or, in terms of data wavefronts, one data wavefront delay.

13.3.1   Analysis of Delay Structure

The initial pipeline structure for examples 2 is shown in Figure 13.15. The upper pipeline consists of three cycles initialized with one DATA wavefront, one NULL wavefront, and one bubble, and the lower pipeline with one cycle initialized to a bubble. Since there are no wavefronts initialized in the lower pipeline, the initialized wavefronts in the upper pipeline form an excess population. There will never be less than two wavefronts in the upper pipeline.

The limiting cycle period for this configuration of the structure is the cycle of the lower pipeline, which is 13 tics. There is one bubble initialized in each pipeline, so the smallest bubble population of the structure is one bubble. Since there are an equal number of bubbles in each pipeline, there is not an excess bubble population, and the two copies of a new bubble must synchronize with each other at the input of the structure. So the bubble renewal path for the wavefront populations is the path with the longest delay, which is this case is the upper acknowledge path. Since there is an excess wavefront population in the upper pipeline, the copy of a newly entered wavefront in the lower pipeline will synchronize with a wavefront

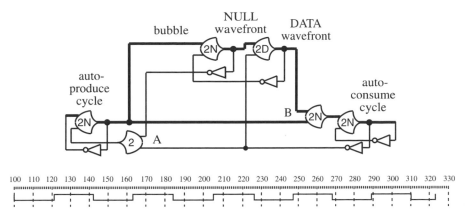

Figure 13.15   Initial pipeline delay structure.

already in the upper pipeline. So the wavefront renewal path for the bubble populations is the data path of the lower pipeline. The renewal paths are shown in Figure 13.16. The critical renewal period is 21 tics. The upper bubble and lower wavefront populations are renewal limited. The throughput of the structure is one wavefront every 21 tics.

Where should one add buffer cycles to enhance the throughput of the structure? It is not immediately obvious. One might first attempt to add a cycle to the lower pipeline, as in Figure 13.17. Adding the cycle to the lower pipeline divides the 13 tic cycle into two cycles, each with a period of 10 tics. So the slowest cycle in the structure is now 10 tics.

However, adding the cycle to the lower pipeline does not change the renewal paths shown in Figure 13.18 nor increase a limiting population. It does, however, increase a renewal period from 21 tics to 24 tics. Because of the one bubble in the upper pipeline there can still only be one wavefront at a time in the lower pipeline. The behavior of the structure remains renewal limited with a throughput of one wavefront every 24 tics, which is a lower throughput than the initial configuration. If more cycles are added to the lower pipeline, the renewal path remains the same, the population period remains the same, and the renewal period increases reducing the throughput of the structure. Clearly, adding buffer cycles to the lower pipeline is not advantageous.

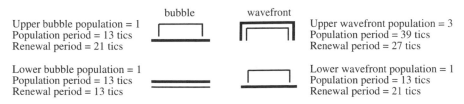

Figure 13.16   Renewal paths for initial configuration of Example 2.

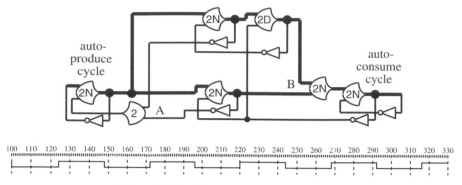

Figure 13.17 Cycle added to lower pipeline.

When a fourth cycle is added to the upper pipeline as in Figure 13.19, the renewal paths change. The upper pipeline now has an excess population of bubbles, which means that the bubble renewal path for the wavefront populations becomes the acknowledge path of the lower pipeline and the renewal paths changes as shown in Figure 13.20. All population periods are now greater than or equal to their renewal periods. The behavior of the structure becomes delay limited and flows at its maximum rate of one wavefront every 13 tics.

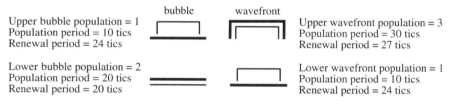

Upper bubble population = 1
Population period = 10 tics
Renewal period = 24 tics

Upper wavefront population = 3
Population period = 30 tics
Renewal period = 27 tics

Lower bubble population = 2
Population period = 20 tics
Renewal period = 20 tics

Lower wavefront population = 1
Population period = 10 tics
Renewal period = 24 tics

Figure 13.18 Renewal paths with 2 cycles in lower pipeline.

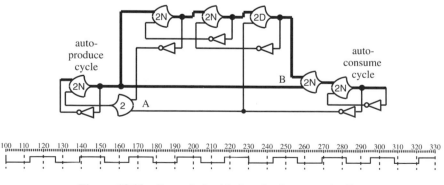

Figure 13.19 Example 2 with 4 cycles in upper pipeline.

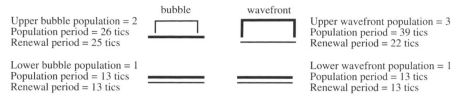

Upper bubble population = 2
Population period = 26 tics
Renewal period = 25 tics

Upper wavefront population = 3
Population period = 39 tics
Renewal period = 22 tics

Lower bubble population = 1
Population period = 13 tics
Renewal period = 13 tics

Lower wavefront population = 1
Population period = 13 tics
Renewal period = 13 tics

Figure 13.20 Renewal path for configuration with 4 cycles in the upper pipeline.

As more buffer cycles are added through the ninth cycle to the upper pipeline, there is no effect whatever on the populations of the lower pipeline and on the throughput of the structure. The upper pipeline bubble population period grows faster than the renewal period. The wavefront renewal period, however, creeps up to the wavefront population period and with the tenth cycle, the upper data path accumulates enough delay that the renewal period for the upper pipeline wavefronts is greater than the population period. Although the wavefront population of the upper pipeline is an excess population and will never be depleted, the renewal wavefronts can lag behind the population flow and cause a wait. The renewal paths are shown in Figure 13.21.

When the tenth cycle is added to the upper pipeline, as shown in Figure 13.22, one more operator with 3 tics of delay is added to the renewal path of Figure 13.23, and the renewal period for the upper wavefront population becomes 40 tics, which exceeds the population period of 39 tics. The throughput of the structure becomes three wavefronts every 40 tics.

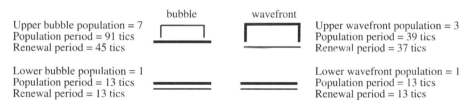

Upper bubble population = 7
Population period = 91 tics
Renewal period = 45 tics

Upper wavefront population = 3
Population period = 39 tics
Renewal period = 37 tics

Lower bubble population = 1
Population period = 13 tics
Renewal period = 13 tics

Lower wavefront population = 1
Population period = 13 tics
Renewal period = 13 tics

Figure 13.21 Renewal paths for Example 2 with 9 cycles in upper pipeline.

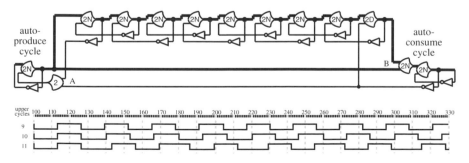

Figure 13.22 Example 2 with 10 cycles in upper pipeline.

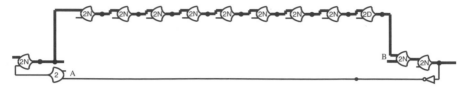

Figure 13.23 Wavefront renewal path for 10 cycle configuration.

As more cycles are added, the wavefront renewal period increases and the throughput decreases. The signal waveforms in Figure 13.22 show the transition from delay-limited behavior with 9 cycles to renewal limited behavior with 10 cycles and 11 cycles in the upper pipeline.

Table 13.2 shows the behavior profile for Example 2. Figure 13.24 shows the waveforms for each of the configurations of Example 2, and Figure 13.25 charts the throughput of the configurations for Example 2.

13.3.2 Summary of Example 2

Again, the example has been shown at its extremes to elucidate its range of behavior. A practical analysis would have ended when the fourth cycle was added to the upper pipeline and the structure began delivering its maximal throughput with the minimal resources.

13.4 EXAMPLE 3: REDUCING THE PERIOD OF THE SLOWEST CYCLE

The throughput of the delay line structure can be further increased by reducing the period of the slowest cycle. But one cannot simply reduce the cycle period.

TABLE 13.2 Example 2 Behavior Profile

Cycles in Upper Pipeline	Limiting Cycle Period	Relevant Population Period	Relevant Renewal Period	Throughput Wave/Period	Throughput Wave/ 100 tics	Limiting Behavior Mode
3	13	13	21	1/21	4.76	Renewal
4	13			1/13	7.69	Delay
5	13			1/13	7.69	Delay
6	13			1/13	7.69	Delay
7	13			1/13	7.69	Delay
8	13			1/13	7.69	Delay
9	13			1/13	7.69	Delay
10	13	39	40	3/40	7.50	Renewal
11	13	39	43	3/43	6.97	Renewal
12	13	39	46	3/46	6.52	Renewal
13	13	39	49	3/49	6.12	Renewal

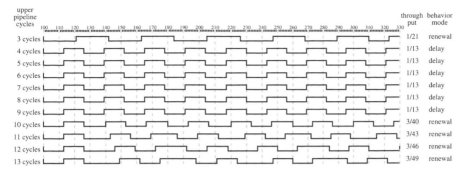

Figure 13.24   Waveforms for each configuration of Example 1.

Changing a cycle period changes the behavior of the structure, and the structure must be reanalyzed.

13.4.1 Finer Grained Pipelining

One way to reduce a cycle period is by further pipelining a slow cycle. After dividing the slow cycle of the lower pipeline into two cycles, the period of each cycle becomes 10 tics. This configuration was shown in Figure 13.17, and it was established that adding more cycles to the lower pipeline did not increase the throughput of the structure. But it does reduce the slowest cycle to a period of 10 tics. So, for this example, we will leave the cycle in the lower pipeline as cycles are added to the upper pipeline. Adding the fourth cycle in the upper pipeline is shown is Figure 13.26. There is no excess bubble population. So when a bubble enters the structure, the two copies have to synchronize to flow out of the system. This means that the bubble renewal path for the wavefront populations is the longest delay path, which is the upper pipeline acknowledge path. Since there are excess wavefronts

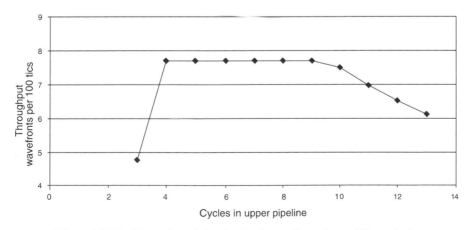

Figure 13.25   Throughput behavior for the configurations of Example 2.

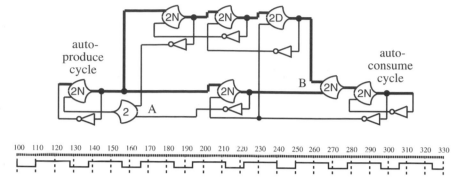

auto-
produce
cycle

auto-
consume
cycle

B

A

Figure 13.26   Four cycles in upper pipeline of Example 3.

in the upper pipeline, the wavefront renewal path for the bubble populations will be
the data path of the lower pipeline. The renewal paths are shown in Figure 13.27.
The upper bubble population and the lower wavefront population are renewal limited, allowing a throughput of two wavefronts every 28 tics.

When the fifth cycle is added to the upper pipeline, the upper pipeline gains an
excess bubble population and the bubble renewal path becomes the acknowledge
path of the lower pipeline, and the renewal paths shown in Figure 13.28 change.
The bubble population of the upper pipeline remains renewal limited. Three bubbles
flow in 32 tics allowing a throughput of three wavefronts every 32 tics. The relevant
renewal path with the signal trace is shown in Figure 13.29.

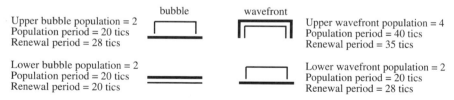

bubble   wavefront

Upper bubble population = 2
Population period = 20 tics
Renewal period = 28 tics

Upper wavefront population = 4
Population period = 40 tics
Renewal period = 35 tics

Lower bubble population = 2
Population period = 20 tics
Renewal period = 20 tics

Lower wavefront population = 2
Population period = 20 tics
Renewal period = 28 tics

Figure 13.27   Renewal paths with four cycles in the upper pipeline.

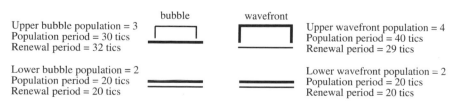

bubble   wavefront

Upper bubble population = 3
Population period = 30 tics
Renewal period = 32 tics

Upper wavefront population = 4
Population period = 40 tics
Renewal period = 29 tics

Lower bubble population = 2
Population period = 20 tics
Renewal period = 20 tics

Lower wavefront population = 2
Population period = 20 tics
Renewal period = 20 tics

Figure 13.28   Renewal paths with five cycles in the upper pipeline.

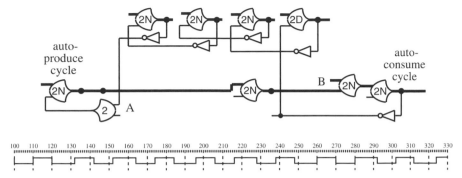

Figure 13.29   Limiting renewal path for five cycles in the upper pipeline.

A sixth cycle must be added to the upper pipeline, as shown in Figure 13.30, to finally deliver the maximum throughput for the structure. Table 13.3 summarizes the behavior of the structure configurations.

13.4.2   Optimizing the Logic

Another approach to reducing the period of the slowest cycle is to optimize the logic. In this case the synchronizing operators A and B can be merged with the regulating operators as shown in Figure 13.31. The period of the cycle in the lower pipeline become 7 tics. All cycles in the structure now have the same cycle period of 7 tics and buffer cycles will have a period of 7 tics so there is no slowest cycle in the structure. The renewal paths are shown in Figure 13.32. The structure is renewal limited to one wave-front every 15 tics.

When the fourth cycle is added, the upper pipeline acquires an excess bubble population and the renewal paths change, as shown in Figure 13.33. The structure is now limited by the renewal of the upper bubble population, which allows a throughput of two wavefronts every 19 tics.

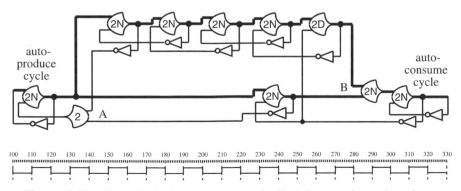

Figure 13.30   Six cycles in the upper pipeline finally deliver maximum throughput.

TABLE 13.3   Example 3 with Two Cycles in Lower Pipeline

Cycles in Upper Pipeline	Limiting Cycle Period	Relevant Population Period	Relevant Renewal Period	Throughput Wave/Period	Throughput Wave/ 100 tics	Limiting Behavior Mode
3	10	10	24	1/24	4.16	Renewal
4	10	20	28	2/28	7.14	Renewal
5	10	30	32	3/32	9.37	Renewal
6	10			1/10	10.0	Delay
7	10			1/10	10.0	Delay
8	10			1/10	10.0	Delay
9	10	40	41	4/41	9.75	Renewal
10	10	40	44	4/44	9.09	Renewal

When the fifth cycle is added, the throughout increases to three wavefronts every 23 tics. The renewal paths are shown in Figure 13.34.

When the sixth cycle is added, the renewal paths of Figure 13.35 show that the upper bubble population is no longer renewal limited, but the upper wavefront population is now renewal limited allowing three wavefronts every 22 tics. As more cycles are added, the renewal path for the upper wavefront population becomes longer and the throughput decreases.

Table 13.4 shows the behavior profile for the structure through adding ten cycles to the upper pipeline. Figure 13.36 shows the throughput graph of the last example superimposed with the performance graphs of the examples with slowest cycles of 13 tics and 10 tics.

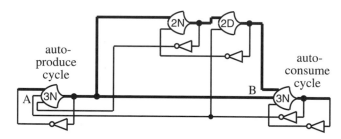

Figure 13.31   Example 3 having no slowest cycle period.

bubble   wavefront

Upper bubble population = 1
Population period = 7 tics
Renewal period = 15 tics

Upper wavefront population = 3
Population period = 21 tics
Renewal period = 21 tics

Lower bubble population = 1
Population period = 7 tics
Renewal period = 7 tics

Lower wavefront population = 1
Population period = 7 tics
Renewal period = 15 tics

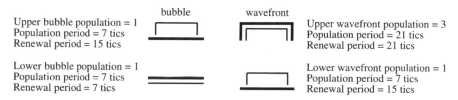

Figure 13.32   Renewals paths with three cycles in upper pipeline.

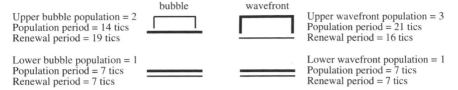

Upper bubble population = 2
Population period = 14 tics
Renewal period = 19 tics

Upper wavefront population = 3
Population period = 21 tics
Renewal period = 16 tics

Lower bubble population = 1
Population period = 7 tics
Renewal period = 7 tics

Lower wavefront population = 1
Population period = 7 tics
Renewal period = 7 tics

Figure 13.33   Renewal paths with four cycles in upper pipeline.

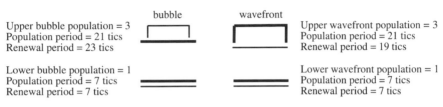

Upper bubble population = 3
Population period = 21 tics
Renewal period = 23 tics

Upper wavefront population = 3
Population period = 21 tics
Renewal period = 19 tics

Lower bubble population = 1
Population period = 7 tics
Renewal period = 7 tics

Lower wavefront population = 1
Population period = 7 tics
Renewal period = 7 tics

Figure 13.34   Renewal paths with five cycles in upper pipeline

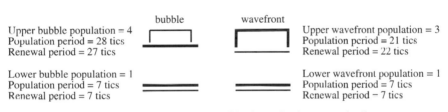

Upper bubble population = 4
Population period = 28 tics
Renewal period = 27 tics

Upper wavefront population = 3
Population period = 21 tics
Renewal period = 22 tics

Lower bubble population = 1
Population period = 7 tics
Renewal period = 7 tics

Lower wavefront population = 1
Population period = 7 tics
Renewal period − 7 tics

Figure 13.35   Renewal paths with six cycles in upper pipeline.

TABLE 13.4   Example 3 with Two Cycles in Lower Pipeline

Cycles in Upper Pipeline	Limiting Cycle Period	Relevant Population Period	Relevant Renewal Period	Throughput Wave/Period	Throughput Wave/ 100 tics	Limiting Behavior Mode
3	7	7	15	1/15	6.66	Renewal
4	7	14	19	2/19	10.5	Renewal
5	7	21	23	3/23	13.0	Renewal
6	7	21	22	3/22	13.6	Renewal
7	7	21	25	3/25	12.0	Renewal
8	7	21	28	3/28	10.7	Renewal
9	7	21	31	3/31	9.68	Renewal
10	7	21	34	3/34	8.82	Renewal

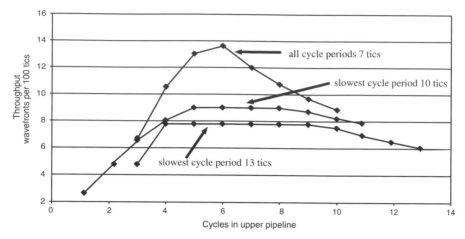

Figure 13.36 Throughput for Example 3 configurations.

When all the cycles have the same period and there is no slowest cycle, there is no throughput plateau. The throughput plateau is formed in relation to the difference in period between the slowest cycles and the rest of the cycles of the structure. If the difference is large, then the plateau will be wide. If the difference is small, then the plateau will be narrow. If there is no difference, there will be no plateau.

The behavior of the last example peaks with one renewal-limited behavior and then transitions to another renewal-limited behavior without achieving the theoretical maximum behavior. The integral nature of pipelines is encountered. Pipelines are structures of whole cycles, and data flows in whole wavefronts. There cannot be fractional cycles, and there cannot be fractional wavefronts. Once cycle configurations are established, altering the cycle structure changes the throughput behavior in discrete jumps. On the graph there is behavior only at each configuration point. Although the points are connected by a line to emphasize their relationships, *there is no behavior between the points*. While it is theoretically possible, with property tuning of delays of the cycles, for one of the behavior points to reach one wavefront every 7 tics or 14.29 wavefronts every 100 tics. It is highly unlikely for practical designs and very likely not worth the cost of the effort.

13.5 EXERCISES

13.1. Explain what happens when wavefronts are initialized in both pipelines. Why can this configuration be ignored in analyzing throughput behavior.

13.2. Explain what happens when cycles are added equally to both pipelines.

13.3. Define a procedure to automatically analyze and optimize two pipeline structures given the cycle structure and the delay data.

CHAPTER 14

Complex Pipeline Structures

Complex structures of pipelines can be built by grafting pipelines one at a time onto a growing structure. Each new pipeline graft can be modeled as a two-pipeline structure, with one pipeline being integral to the existing structure and the other pipeline being the one grafted. The pipeline integral to the existing structure will have already been integrated into the structure, and modifying it could corrupt the integrity of the existing structure. So in most cases the integral pipeline will be taken as a static referent pipeline and the pipeline being grafted will be the one that is modified.

14.1 LINEAR FEEDBACK SHIFT REGISTER EXAMPLE

The example will be a fairly complex linear feedback shift register (LFSR). Figure 14.1 shows the model of the LFSR, which is a ring structure with XOR functions crossing through it. The squares are registers initialized to the value indicated. The data flow clockwise around the pipeline ring, with pipelines branching across the ring through the XOR functions resulting in a complex structure of data paths.

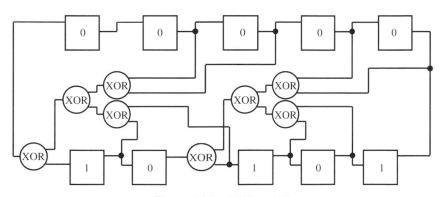

Figure 14.1 LFSR model.

Logically Determined Design: Clockless System Design with NULL Convention Logic[TM], by Karl M. Fant
ISBN 0-471-68478-3 Copyright © 2005 John Wiley & Sons, Inc.

The discussion will begin with a mapping of the model into a collapsed data path NCL structure. The XOR combinational functions will be represented by 3 of 3 operators initialized to NULL (3N) and identified with XOR. Fan-ins of data paths and acknowledges will be merged with the regulator operators so that all the cycles in the structure will have a period of 7 tics. There will be no slowest cycle in the structure. This will provide a more thorough demonstration of the issues involved in design for throughput. Because the structure is storing wavefronts in the pipelines and has almost no logic, it involves lots of pipeline cycles. The reader should overlook this for the time being. While it is not a very good example of efficient implementation, it is a very good example for pipeline structure synthesis.

An initial mapping of the LFSR into NCL, as shown in Figure 14.2, maps each register into four cycles with an initialized DATA wavefront, an initialized NULL wavefront, and two bubbles. The XOR functions are also pipelined, so the complete structure is gate level pipelined.

Is this an optimal structure? Can one tell? Given that all the cycle periods are 7 tics, how closely can a throughput for the structure of one wavefront every 7 tics be approached? Can the throughput be improved by adding or subtracting buffer cycles at appropriate places? How can one tell where to add or subtract buffer cycles? Given the dynamic nature of all the interrelationships of the structure, trying to answer these questions might seem quite daunting, but there is a simple methodology that provides precise answers to all these questions.

14.2 GRAFTING PIPELINES

The basic procedure is to break the structure into component pipelines, choose a core ring or a core two pipeline structure, and then by grafting pipelines onto this core, rebuild the structure one pipeline at a time. Each graft can be characterized as an isolated two-pipeline structure and analyzed exactly as the examples in the previous chapter were analyzed. It is convenient if the core is chosen to contain the slowest cycle in the structure, which will establish an initial throughput limit for the structure.

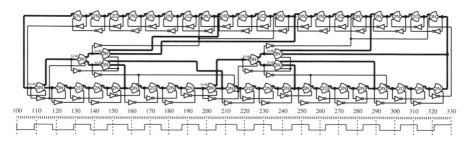

Figure 14.2 NCL collapsed data path model of LFSR.

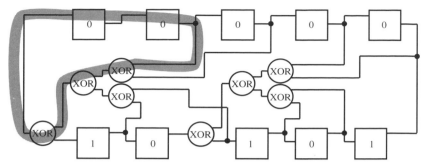

Figure 14.3 Core ring component of the LFSR.

14.2.1 Step 1

For the present example there is no slowest cycle. A core ring is chosen and the larger structure will be rebuilt from this ring. Figure 14.3 shows the LFSR model and the ring chosen as the core. The corresponding NCL circuit shown in Figure 14.4 is isolated from the initial NCL structure of Figure 14.2. Available inputs to operators are extended to substitute for the inputs from the rest of the structure that are not yet present. For instance, a single input from the structure is presented to both inputs of each XOR operator. The substituted inputs will receive their appropriate inputs as component pipelines are grafted.

The first task is to optimize the throughput of the ring. There are 11 cycles with 4 wavefronts (2 DATA and 2 NULL) and 7 bubbles initialized in the ring. Building the behavior profile in Table 14.1, it can be seen that 9 cycles in the ring provides the optimal throughput. So 2 cycles can be removed from the ring. The XOR cycles cannot be removed and the wavefront initializing cycles cannot be removed. The cycles that can be removed are the buffer or bubble cycles in the upper rank of

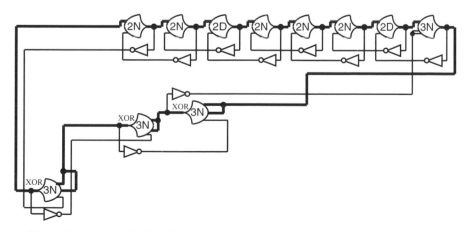

Figure 14.4 Step 1 initial NCL model and isolation model of core component ring.

TABLE 14.1 Step 1 Ring Behavior Profile

Cycles in Ring	Limiting Cycle Period	Bubbles in Ring	Bubble Population Period	Bubble Rejoin Period	Wavefronts in Ring	Wavefront Population Period	Wavefront Rejoin Period	Throughput Waves/Period	Throughput Waves/100 tics	Limiting Behavior Mode
14	7	10	70	56	4	28	42	4/42	9.52	Wavefront
13	7	9	63	52	4	28	39	4/39	10.26	Wavefront
12	7	8	56	48	4	28	36	4/36	11.11	Wavefront
11	7	7	49	44	4	28	33	4/33	12.12	Wavefront
10	7	6	42	40	4	28	30	4/30	13.33	Wavefront
9	7	5	35	36	4	28	27	5/36	13.88	Bubble
8	7	4	28	32	4	28	24	4/32	12.5	Bubble
7	7	3	21	28	4	28	21	3/28	10.7	Bubble
6	7	2	14	24	4	28	18	2/24	8.33	Bubble
5	7	1	7	20	4	28	15	1/20	5.0	Bubble

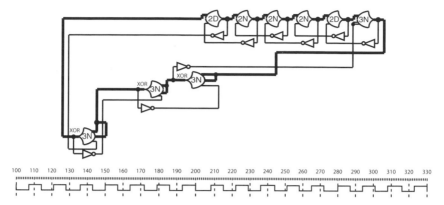

Figure 14.5 Step 1 optimized core ring component.

cycles. Figure 14.5 show the optimized core ring with 9 cycles. The throughput is five wavefronts every 36 tics (13.88 wavefronts per 100 tics) with the repeating pattern of wavefront periods 7, 7, 7, 7, 8. It is already determined that a throughput of one wavefront every 7 tics cannot be achieved.

14.2.2 Step 2

In step 2 the first component pipeline is grafted onto the core structure. Figure 14.6 shows the graft component on the model. Figure 14.7 shows the pipeline segment of the initial NCL model grafted to the ring. The new graft pipeline is inserted on the existing structure, in this case the ring, with a fan-out at the beginning of the graft pipeline and a fan-in at the end of the graft pipeline. A two-pipeline NCL isolation model, shown in Figure 14.8, is constructed from the graft pipeline and the segment of the growing structure involved in the graft. The isolation model models the two-pipeline structure between these two insertion/synchronization points. An auto-produce cycle provides input and an auto-consume cycle provides output. The pipeline from the existing structure will be shown as the lower pipeline, and the pipeline being grafted as the upper pipeline.

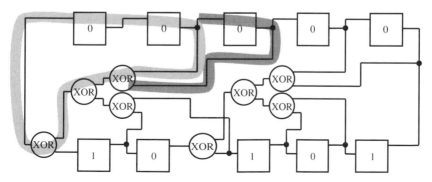

Figure 14.6 Step 2 graft model.

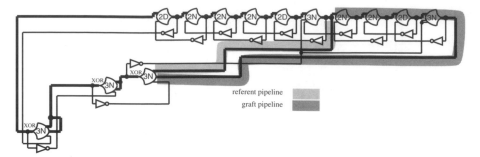

Figure 14.7 Initial NCL model for the first pipeline graft onto the core ring component.

The lower pipeline has already been subject to optimization and cannot be altered, so it becomes the referent pipeline. The limiting throughput for the isolation model (slowest cycle equivalent) is the throughput of the existing structure that the lower pipeline was taken from. The upper pipeline is the variable pipeline that is to be modified. The idea is to modify the upper pipeline to support the throughput of the existing structure it is being grafted onto. If this can be achieved, then the new structure with the new graft will maintain its throughput. If this cannot be achieved, then the new graft structure will become the limiting factor, reducing the throughput of the whole structure and establishing a new throughput limit for the structure. The throughput of a structure is limited by the throughput of its slowest component structure, just as the throughput of a pipeline is limited by the throughput of its slowest cycle.

The behavior profile of the isolation model, shown in Table 14.2, which is identical to the final version of Example 3 in the previous chapter, is built to determine the optimal configuration for the graft. The peak throughput of 13.64 wavefronts per 100 tics for the isolation structure occurs with 6 cycles in the upper pipeline. With the given operators and the discrete behavior of pipeline structures, this cannot be improved. This is less than the peak throughput for the ring, which means that the graft will establish a new throughput limit for the structure.

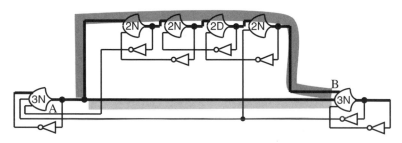

Figure 14.8 Isolation model for the step 2 graft.

TABLE 14.2   Step 2 Isolation Structure Behavior Profile

Cycles in Upper Pipeline	Limiting Cycle Period	Relevant Population Period	Relevant Renewal Period	Throughput Wave/ Period	Throughput Wave/100 tics	Limiting Behavior Mode
3	7	7	15	1/15	6.66	Renewal
4	7	14	19	2/19	10.52	Renewal
5	7	21	23	3/23	13.04	Renewal
6	7	21	22	3/22	13.64	Renewal
7	7	21	25	3/25	12.00	Renewal
8	7	21	28	3/28	10.71	Renewal
9	7	21	31	3/31	9.68	Renewal
10	7	21	34	3/34	8.82	Renewal

The initial graft pipeline had five cycles. It can be extended to six cycles by adding a buffer cycle. Figure 14.9 shows the new structure with the new 6 cycle pipeline grafted. The throughput is 3 wavefronts every 22 tics (13.64 wavefronts per 100 tics) with the repeating pattern of wavefront periods 7, 7, and 8.

Since in this case the core is a ring structure, every new pipeline grafted onto the core also forms a new ring. The behavior of this new ring must be considered as well as the behavior of the local structure. The new ring of step 2, shown in Figure 14.9, contains 14 cycles with six initialized wavefronts leaving eight bubbles in the ring. Table 14.3 shows the ring to be perfectly balanced, with the population periods equal to the rejoin periods. The new ring does not limit the throughput of the new structure.

14.2.3 Step 3

Step 3 is a fairly large graft structure, which is shown on the model in Figure 14.10. The initial NCL graft model is shown in Figure 14.11. The two-pipeline isolation model is shown in Figure 14.12.

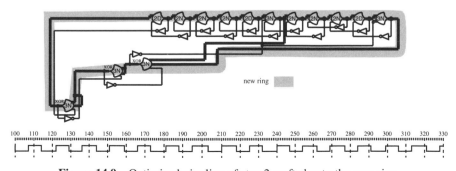

Figure 14.9   Optimized pipeline of step 2 grafted onto the core ring.

TABLE 14.3   Performance Parameters for the Graft Ring of Step 2

		Population Period	Rejoin Period	Limiting Behavior Mode
Cycles	14			
Wavefronts	6	42	42	Delay
Bubbles	8	56	56	Delay

Table 14.4 shows the behavior profile for the two pipeline isolation model structure. It shows that with 11 cycles the structure will deliver a throughput of 14.00 wavefronts per 100 tics. This is well beyond the performance of the existing structure of 13.64 wavefronts per 100 tics. So the pipeline can be grafted to the existing structure with two fewer cycles than the initial model. A question arises about where to remove the two cycles. Part of the graft is in the upper part of the LFSR ring, and part of the graft is in the lower part of the LFSR ring. Cycles can be removed from either of these two positions. Currently it does not make any difference where the cycles are removed, but it might make a difference later. At this point one cycle will be removed from each position, and it will be noted that this might be a flexibility in relation to later grafts. Figure 14.13 shows the optimized pipeline grafted onto the growing structure. The throughput of the structure remains 3 wavefronts every 22 tics (13.64 wavefronts per 100 tics).

The new ring of step 3, shown in Figure 14.13, contains 23 cycles with ten initialized wavefronts leaving thirteen bubbles in the ring. Table 14.5 shows the ring to be slightly bubble limited to 13 wavefronts every 92 tics or 14.13 wavefronts per 100 tics. This is a greater throughput than the current limit of 13.64 wavefronts per 100 tics, so the new ring does not limit the throughput of the new structure.

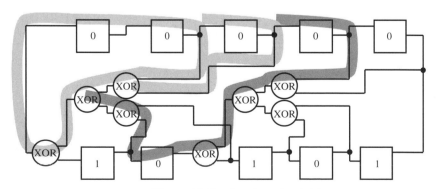

Figure 14.10   Step 3 graft model.

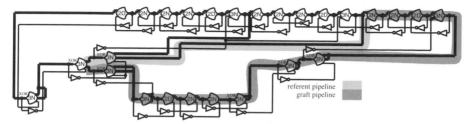

Figure 14.11   Step 3 initial NCL graft model.

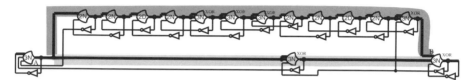

Figure 14.12   Isolation model for step 3 graft.

14.2.4   Step 4

Step 4 is identical to step 2. The isolation structure is identical and the optimization is identical. Figure 14.14 shows the step 4 graft model and Figure 14.15 the step 4 pipeline component grafted onto the structure. The throughput of the structure remains 3 wavefronts every 22 tics (13.64 wavefronts per 100 tics).

The new ring of step 4, shown in Figure 14.15, contains 28 cycles with 12 initialized wavefronts leaving 16 bubbles in the ring. Table 14.6 shows the ring to be perfectly balanced with the population period equal to the renewal period. The new ring does not limit the throughput of the new structure.

TABLE 14.4   Step 3 Isolation Structure Behavior Profile

Cycles in Upper Pipeline	Limiting Cycle Period	Relevant Population Period	Relevant Renewal Period	Throughput Wave/ Period	Throughput Wave/100 tics	Limiting Behavior Mode
7	7	21	34	3/34	8.82	Renewal
8	7	28	38	4/38	10.53	Renewal
9	7	35	42	5/42	11.90	Renewal
10	7	42	46	6/46	13.04	Renewal
11	7	49	50	7/50	14.00	Renewal
12	7	42	44	6/44	13.63	Renewal
13	7	42	47	6/47	12.76	Renewal

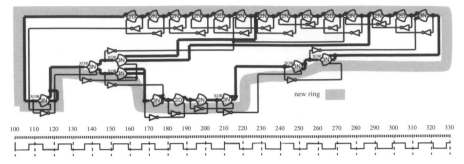

Figure 14.13 Optimized pipeline of step 3 grafted onto the growing structure.

TABLE 14.5   Performance Parameters for the Graft Ring of Step 3

		Population Period	Rejoin Period	Limiting Behavior Mode
Cycles	23			
Wavefronts	10	70	69	Delay
Bubbles	13	91	92	Bubble

14.2.5   Step 5

The graft model for step 5 is shown in Figure 14.16. The initial NCL graft model is shown in Figure 14.17. The isolation model is shown in Figure 14.18.

Table 14.7 shows the behavior profile for the step 5 isolation model. Seven cycles give the optimal throughput for the structure. The other configurations offers less

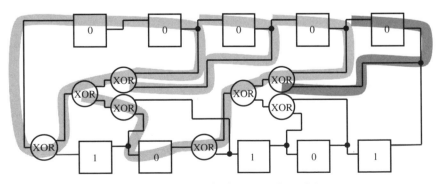

Figure 14.14   Step 4 graft model.

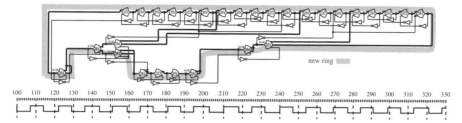

Figure 14.15 Step 4 pipeline grafted onto structure.

TABLE 14.6 Performance Parameters for the Graft Ring of Step 4

		Population Period	Rejoin Period	Limiting Behavior Mode
Cycles	28			
Wavefronts	12	84	84	Delay
Bubbles	16	112	112	Delay

throughput than the 13.64 wavefronts per 100 tics that is the current limit for the structure. The grafted pipeline with 7 cycles is shown in Figure 14.19. The throughput of the structure remains 3 wavefronts every 22 tics (13.64 wavefronts per 100 tics).

The new ring of step 5, shown in Figure 14.19, contains 33 cycles with 14 initialized wavefronts leaving 19 bubbles in the ring. Table 14.8 shows the ring to be slightly wavefront limited to 14 wavefronts every 99 tics or 14.14 wavefronts per 100 tics. This is a greater throughput than the current limit of 13.64 wavefronts per 100 tics, so the new ring does not limit the throughput of the new structure.

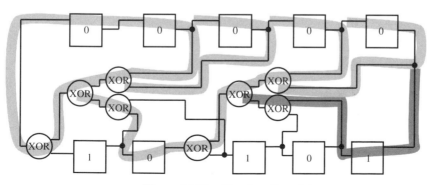

Figure 14.16 Step 5 graft model.

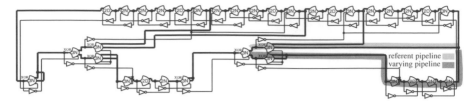

Figure 14.17 Step 5 initial NCL graft model.

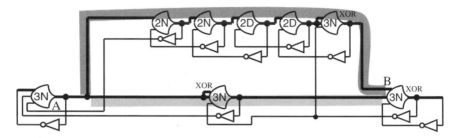

Figure 14.18 Step 5 isolation model.

TABLE 14.7 Step 5 Isolation Structure Behavior Profile

Cycles in Upper Pipeline	Limiting Cycle Period	Relevant Population Period	Relevant Renewal Period	Throughput Wave/ Period	Throughput Wave/100 tics	Limiting Behavior Mode
3	7	10	18	1/18	4.16	Renewal
4	7	20	22	2/22	9.09	Renewal
5	7	30	26	3/26	11.54	Renewal
6	7	40	30	4/30	13.33	Renewal
7	7	40	29	4/29	13.79	Renewal
8	7	40	32	4/32	12.50	Renewal
9	7	40	35	4/35	11.43	Renewal

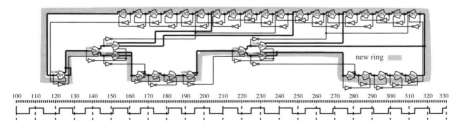

Figure 14.19 Step 5 pipeline grafted onto structure.

TABLE 14.8   Performance Parameters for the Graft Ring of Step 5

		Population Period	Rejoin Period	Limiting Behavior Mode
Cycles	33			
Wavefronts	14	98	99	14
Bubbles	19	133	132	19

14.2.6   Step 6

Step 6 is identical to steps 2 and 4. The isolation structure is identical and the optimization is identical. Figure 14.20 shows the step 6 graft model, and Figure 14.21 the step 6 pipeline component grafted onto the structure. The resulting structure, however, does not maintain the expected throughput. The simulated throughput is 8 wavefronts in 59 tics (13.56 wavefronts per 100 tics) with the repeating sequence of wavefront periods 7, 7, 8, 7, 7, 8, 7, 8.

With graft 6 there are 4 inputs presented to the EOR structure. These inputs do not all arrive simultaneously. There is slight jitter in their arrival, and occasionally there is a one tic difference in their arrival. What is occurring here is another consequence of the integer nature of the structure and its behavior. Just as there is no way in the context of the available components to adjust the throughput of the ring of step 1 to 1 wavefront every 7 tics, there is no way to adjust the arrivals of the wavefronts at the EOR gates by one tic with the given components. The resulting throughput establishes a new limit for the structure.

While this jitter can probably be modeled and analyzed in a similar manner, a methodology has not been developed and is not presented here. In this case the simulation results are simply accepted.

The new ring of step 6, shown in Figure 14.21, contains 38 cycles with sixteen initialized wavefronts leaving 22 bubbles in the ring. Table 14.9 shows the ring to

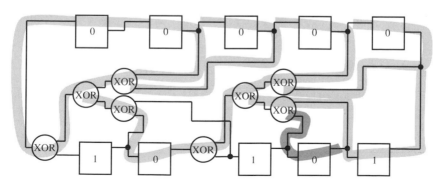

Figure 14.20   Step 6 graft model.

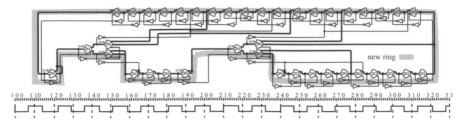

Figure 14.21 Step 6 pipeline grafted onto structure.

be slightly wavefront limited to 16 wavefronts every 114 tics or 14.03 wavefronts per 100 tics. This is a greater throughput than the current limit of 13.56 wavefronts per 100 tics, so the new ring does not limit the throughput of the new structure.

TABLE 14.9   Performance Parameters for the Graft Ring of Step 6

		Population Period	Rejoin Period	Limiting Behavior Mode
Cycles	38			
Wavefronts	16	112	114	Wavefront
Bubbles	22	154	152	Delay

14.2.7   Step 7

The graft model for step 7 is shown in Figure 14.22. The initial NCL model is shown in Figure 14.23. The isolation model is shown in Figure 14.24.

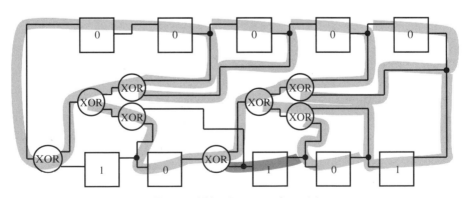

Figure 14.22 Step 7 graft model.

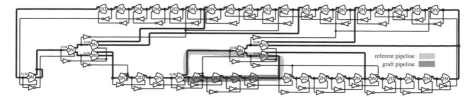

Figure 14.23   Step 7 initial NCL graft model.

Table 14.10 shows the behavior profile for the step 7 isolation model. Eight cycles give the optimal throughput for the structure. The other configurations offer less throughput than the 13.56 wavefronts per 100 tics that is the current limit for the structure. The grafted pipeline with 8 cycles is shown in Figure 14.25. The simulated throughput remains 8 wavefronts every 59 tics (13.56 wavefronts per 100 tics).

The new ring of step 7, shown in Figure 14.25, contains 43 cycles with 18 initialized wavefronts leaving 25 bubbles in the ring. Table 14.11 shows the ring to be slightly wavefront limited to 18 wavefronts every 129 tics or 13.95 wavefronts per

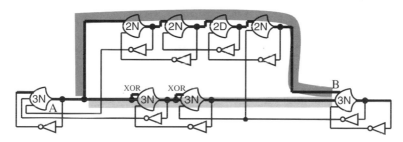

Figure 14.24   Step 7 isolation model.

TABLE 14.10   Step 7 Isolation Model Behavior Profile

Cycles in Upper Pipeline	Limiting Cycle Period	Relevant Population Period	Relevant Renewal Period	Throughput Wave/ Period	Throughput Wave/100 tics	Limiting Behavior Mode
3	7	7	21	1/21	4.76	Renewal
4	7	14	25	2/25	8.00	Renewal
5	7	21	29	3/29	10.34	Renewal
6	7	28	33	4/33	12.12	Renewal
7	7	35	37	5/37	13.51	Renewal
8	7	35	36	5/36	13.89	Renewal
9	7	35	39	5/39	12.82	Renewal
10	7	35	42	5/42	11.90	Renewal

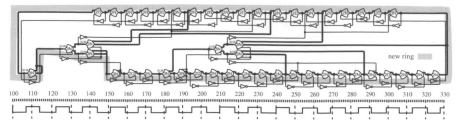

Figure 14.25 Step 7 pipeline grafted onto structure.

TABLE 14.11 Performance Parameters for the Graft Ring of Step 7

		Population Period	Rejoin Period	Limiting Behavior Mode
Cycles	43			
Wavefronts	18	126	129	Wavefront
Bubbles	25	175	172	Delay

100 tics. This is a greater throughput than the current limit of 13.56 wavefronts per 100 tics, so the new ring does not limit the throughput of the new structure.

14.2.8 Step 8

Step 8 is identical to step 7. The isolation structure is identical and the optimization is identical. Figure 14.26 shows the step 8 graft model and Figure 14.27 the step 8 pipeline component grafted onto the structure. The simulated throughput remains 8 wavefronts every 59 tics.

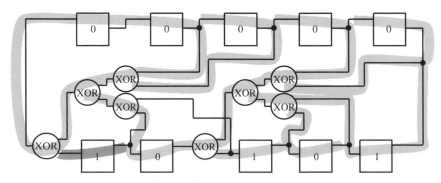

Figure 14.26 Step 8 graft model.

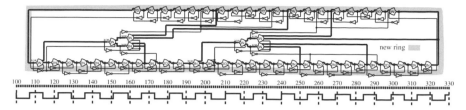

Figure 14.27   Step 8 pipeline grafted onto structure.

The new ring of step 8, shown in Figure 14.27, contains 48 cycles with 20 initialized wavefronts, leaving 28 bubbles in the ring. Table 14.12 shows the ring to be slightly wavefront limited to 20 wavefronts every 144 tics or 13.88 wavefronts per 100 tics. This is a greater throughput than the current limit of 13.56 wavefronts per 100 tics, so the new ring does not limit the throughput of the new structure.

TABLE 14.12   Performance Parameters for the Graft Ring of Step 8

		Population Period	Rejoin Period	Limiting Behavior Mode
Cycles	48			
Wavefronts	20	140	144	Wavefront
Bubbles	28	196	192	Delay

14.2.9   Step 9

The model for the last graft is shown in Figure 14.28. This graft is an internal path, but it still has a beginning fan-out and an ending fan-in and a corresponding structure

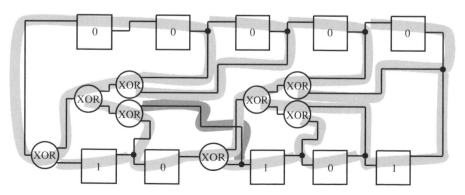

Figure 14.28   Step 9 graft model.

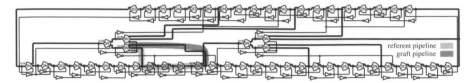

Figure 14.29   Step 9 NCL graft model.

pipeline segment between the fan-out and the fan-in. Figure 14.29 shows the NCL model of the graft. The isolation structure is shown in Figure 14.30.

From the behavior profile of step 2 it can be seen that throughput of the current configuration is 3 wavefronts in 23 tics or 13.04 wavefronts per 100 tics. If cycles are added to the upper pipeline as shown in Table 14.13, the throughput just gets worse.

It will be noticed that the isolation model for step 9 is the isolation model for step 2 upside down and that the throughput of the structure can be increased to 3 wavefronts in 22 tics or 13.64 wavefronts per 100 tics by adding a cycle to the lower referent pipeline. But the referent pipeline is part of the already established structure, and in general, it cannot be modified without compromising an optimization already established.

There are two options. One can live with the reduced throughput of the graft, establishing a new reduced throughput for the structure, or one can find a way to add a cycle to the referent pipeline segment without compromising the overall structure. As it turns out the referent pipeline segment for this graft is part of the step 3

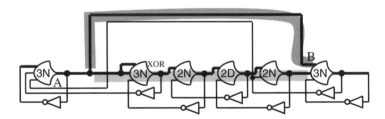

Figure 14.30   Step 9 isolation model.

TABLE 14.13   Step 9 Isolation Model Behavior Profile

Cycles in Upper Pipeline	Limiting Cycle Period	Relevant Population Period	Relevant Renewal Period	Throughput Wave/ Period	Throughput Wave/100 tics	Limiting Behavior Mode
1	7	21	23	3/23	13.04	Bubble
2	7	21	26	3/26	11.54	Bubble
3	7	21	29	3/29	10.34	Bubble
4	7	21	32	3/32	9.38	Bubble

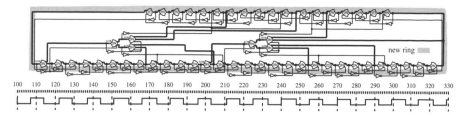

Figure 14.31   Final LFSR structure with step 9 pipeline grafted onto structure.

graft, and it will be recalled that there was some flexibility in step 3 about where cycles could be removed from the graft pipeline. At the time one cycle was removed from each eligible part of the graft pipeline.

It now appears that if both cycles had been removed from the upper segment of the graft pipeline, then there would be one more cycle in the lower segment of the graft, which now is the referent pipeline of the current isolation structure and is just what is needed to provide optimal throughput for the isolation structure. Changing the upper section of the step 3 graft does not compromise another optimized structure so changing the upper section should have no adverse effect. Figure 14.31 shows the final structure with the altered step three to accommodate the optimization for step 9.

Upon adding the fourth input to the EOR structure, input jitter is again encountered and slightly compromises the throughput of the structure. The simulated throughput is 5 wavefronts in 37 tics (13.51 wavefronts per 100 tics) with the repeating sequence of wavefront periods 7, 7, 8, 7, 8.

The new ring of step 9, shown in Figure 14.31, contains 38 cycles with 16 initialized wavefronts leaving 22 bubbles in the ring. Table 14.14 shows the ring to be slightly wavefront limited to 16 wavefronts every 114 tics or 14.03 wavefronts per 100 tics. This is a greater throughput than the current limit of 13.51 wavefronts per 100 tics, so the new ring does not limit the throughput of the new structure.

TABLE 14.14   Performance Parameters for the Graft Ring of Step 9

		Population Period	Rejoin Period	Limiting Behavior Mode
Cycles	38			
Wavefronts	16	112	114	Wavefront
Bubbles	22	154	152	Delay

14.2.10 Summary of Results

All of the questions asked in Section 14.1 are answered. The structure can be modified to determinably optimal performance by adding and removing buffer cycles at specifically determinable places in the structure.

The example began with an LFSR function defined in traditional terms of memory registers and logic functions. It was mapped to a first approximation NCL collapsed data path expression composed of cycles with a throughput of 1 wavefront every 7 tics. This initial NCL expression was decomposed to a core ring which was optimized to a throughput of 5 wavefronts every 36 tics (13.88 wavefronts per 100 tics). The expression was then rebuilt by grafting pipelines one at a time onto the ring. Each graft was optimized for throughput. Over the nine steps of rebuilding the expression, three compromises to the throughput were encountered. The first was in step 2 in which the graft simply had a lower throughput capability of 3 wavefronts every 22 tics (13.64 wavefronts per 100 tics). The other two compromises were the input jitter to the EOR structures in step 6 with 8 wavefronts in 59 tics (13.56 wavefronts per 100 tics) and step 9 with 5 wavefronts in 37 tics (13.51 wavefronts per 100 tics), which became the final throughput of the structure.

Because there is no slowest cycle, and hence no throughput plateau, the resulting structure is uniquely optimal. Every graft except graft 3 was tuned to the only configuration that delivered adequate throughput. The flexibility in graft 3 was used up in optimizing graft 9. If a cycle is added anywhere in the structure, it will decrease the throughput. If a cycle is removed anywhere in the structure, it will decrease the throughput.

Using purely static relationships derived from the structure itself and the delay components of the structure, a complex structure of pipelines was constructed with demonstrably optimal efficiency delivering the maximal throughput performance with minimal resources. The construction began with a ring with a throughput of 5 wavefronts every 36 tics and resulted in a complex structure of pipelines with a throughput of 5 wavefronts every 37 tics.

14.3 THE LFSR WITH A SLOW CYCLE

Adding a slow cycle to the structure imposes a throughput plateau on the behavior of the structure. The structure can then be configured to the end of the plateau, with the fewest cycles thereby reducing the size of the structure. A delay of 4 will be inserted into the data path of the core ring making a slowest cycle with a period of 11 tics.

Table 14.15 is the behavior profile for the step 1 ring with a 4 tic delay inserted in the data path. The delay in the data path increases the population periods and the wavefront rejoin period but has no effect on the bubble rejoin period, since the delay is not in the bubble path of the ring. By comparing Table 14.15 with Table 14.1, it can be seen that for 5 and 6 cycles in the bubble-limited mode the throughput for the two tables is identical. The slow cycle imposes a plateau on the behavior of the structure while the wavefront and bubble-limited behavior beyond the plateau remain the same. The optimal configurations for the slow delay example can be extrapolated directly from the behavior profile tables already compiled above. Table 14.1 shows that a step 1 ring with 7 cycles will support a throughput of 10.7 wavefronts per 100 tics, so the step 1 ring only needs 7 cycles to support a throughput of 9.09 wavefronts per 100 tics. This extrapolation is verified

TABLE 14.15 Step 1 Ring Behavior Profile with an 11 tic Cycle Period

Cycles in Ring	Limiting Cycle Period	Bubbles in Ring	Bubble Population Period	Bubble Rejoin Period	Wavefronts in Ring	Wavefront Population Period	Wavefront Rejoin Period	Throughput Waves/ Period	Throughput Waves/100 tics	Limiting Behavior Mode
14	11	10	110	56	4	44	46	4/47	8.51	Wavefront
13	11	9	99	52	4	44	43	1/11	9.09	Delay
12	11	8	88	48	4	44	40	1/11	9.09	Delay
11	11	7	77	44	4	44	37	1/11	9.09	Delay
10	11	6	66	40	4	44	34	1/11	9.09	Delay
9	11	5	55	36	4	44	31	1/11	9.09	Delay
8	11	4	44	32	4	44	28	1/11	9.09	Delay
7	11	3	33	28	4	44	25	1/11	9.09	Delay
6	11	2	22	24	4	44	22	2/24	8.33	Bubble
5	11	1	11	20	4	44	19	1/20	5.0	Bubble

by Table 14.15. The following is the extrapolation for the slow cycle example from the above behavior profiles for each graft step:

The step 1 initial ring can be 7 cycles instead of 9 cycles.

The step 2 graft can be 4 cycles instead of 6 cycles.

The step 3 graft can be 8 cycles instead of 11 cycles.

The step 4 graft can be 4 cycles instead of 6 cycles.

The step 5 graft can be 4 cycles instead of 7 cycles.

The step 6 graft can be 4 cycles instead of 6 cycles.

The step 7 graft can be 5 cycles instead of 8 cycles.

The step 8 graft can be 5 cycles instead of 8 cycles.

The step 9 graft can be 4 cycles instead of 6 cycles.

Figure 14.32 shows the resulting configuration. This structure (34 operators) is considerably smaller than that of the first example (54 operators).

Simulation, however, reveals a repeating pattern of wavefront periods of 15, 14, 12, 11, 12, 11, 12, 13, 12, 14, 15, 11, 12, 11, 12, 11, 14, 11. This is a throughput of 18 wavefronts every 223 tics or 8.07 wavefronts per 100 tics. This is somewhat less than the expected throughput of 1 wavefront per 11 tics or 9.09 wavefronts per 100 tics. Since slowest cycle will cast its shadows through the structure making all other cycles wait and essentially buffering any jitter in other parts of the structure, it should be possible to achieve a throughput of exactly 1 wavefront every 11 tics.

The next step is to check the rings. Table 14.16 is the behavior profile for each ring of each graft step in the new configuration. It can be seen that for steps 7, 8, and 9 the graft rings do not support the desired throughput. These rings are bubble limited, meaning that there are too few cycles in the ring and that cycles should be added to increase the throughput.

If a cycle is added to the step 7 graft and hence to its ring, it also adds a cycle to the rings of step 8 and step 9. Table 14.17 shows the result of this added cycle. The ring of step 8 is still bubble limited. Adding a cycle to the structure of step 8, and hence to its ring, does not add a cycle to any other ring in the structure. The result of this addition is shown in Table 14.18.

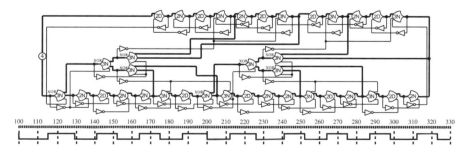

Figure 14.32 Initial configuration for LFSR with slow cycle.

TABLE 14.16 Ring Behavior for Each Step of the LFSR

Step	Cycles	Wavefronts	Wavefront Population Period	Wavefront Rejoin Period	Bubbles	Bubble Population Period	Bubble Rejoin Period	Limiting Behavior Mode	Waves/100 tics
1	7	4	44	22	3	33	28	Delay	9.09
2	10	6	66	30	4	44	40	Delay	9.09
3	17	10	110	51	7	77	68	Delay	9.09
4	20	12	132	60	8	88	80	Delay	9.09
5	23	14	154	69	9	99	92	Delay	9.09
6	26	16	176	78	10	110	104	Delay	9.09
7	28	18	198	87	10	110	112	Bubble	8.92
8	30	20	220	96	10	110	120	Bubble	8.33
9	25	16	176	81	9	99	100	Bubble	9.00

TABLE 14.17 One Cycle Added to the Graft of Step 7

Step	Cycles	Wavefronts	Wavefront Population Period	Wavefront Rejoin Period	Bubbles	Bubble Population Period	Bubble Rejoin Period	Limiting Behavior Mode	Waves/100 tics
7	29	18	198	87	11	121	116	Delay	9.09
8	31	20	220	96	11	121	124	Bubble	8.87
9	26	16	176	81	10	110	104	Delay	9.09

TABLE 14.18 One Cycle Added to the Graft of Step 8

Step	Cycles	Wavefronts	Wavefront Population Period	Wavefront Rejoin Period	Bubbles	Bubble Population Period	Bubble Rejoin Period	Limiting Behavior Mode	Waves/100 tics
7	29	18	198	87	11	121	116	Delay	9.09
8	32	20	220	96	12	132	128	Delay	9.09
9	26	16	176	81	10	110	104	Delay	9.09

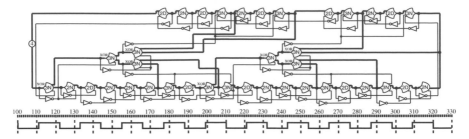

Figure 14.33   Optimal configuration of LFSR with slow cycle.

The graft rings no longer limit the throughput of the structure to less than 9.09 wavefronts per 100 tics. Simulation indicates that the throughput of the structure is indeed exactly 9.09 wavefronts per 100 tics, with one wavefront every 11 tics with no jitter. The final configuration and its waveform is shown in Figure 14.33. This is an optimal configuration in the sense that if any cycle is removed from the structure, the throughput will decrease. Since the slow cycle imposes plateaus on the behavior profiles, a great many cycles can be added to the structure without altering its behavior in any way.

14.4 SUMMARY

Both examples were synthesized to optimal configuration of maximal throughput with minimal resources entirely in terms of static relationships derived from the structure itself and the static delay components of the structure. While a dynamic simulation was used to verify and guide the synthesis, it was not used to search a large possibility space. The possibility space was reduced almost to unity in terms of the static analysis.

The methodology consists of constructing behavior profile tables of static relationships among static parameters and searching the table for a maximum value. The methodology is easily automatable.

14.5 EXERCISES

14.1. Choose a function—ALU, DSP function, instruction decoder, register file, etc.—and construct an optimal NCL configuration. Determine the throughput and the latency by analysis and verify by simulation. Identify and discuss the critical design issues.

14.2. Develop a procedure to analyze and optimize the performance of any two-pipeline structure.

14.3. Define a procedure to automatically compose complex structures of pipelines by grafting in relation to a behavioral specification.

14.4. Develop a methodology to analyze arrival jitter at multiple-input operators.

Logically Determined Wavefront Flow

Wavefronts spontaneously flow through cycles composed with shared completeness paths by virtue of the cycles, themselves, continuously striving to transition. The logical behavior of each completeness path implements a four-phase handshake protocol [35,40] that provides a conveniently intuitive context for talking about the behavior of wavefront flow.

A.1 SYNCHRONIZATION

Figure A.1 reviews the 2NCL the registration stage, which forms the shared completeness path spanning the monotonically transitioning data path. The 2NCL version of the four-phase handshake protocol is shown in Figure A.2 as a polite conversation between registration stages.

A.2 WAVEFRONTS AND BUBBLES

This polite conversation, together with the completeness relationships of the registration stages, results in a forward flow of wavefronts and a backward flow of requests. The backward flowing request is called a bubble. Wavefronts flow forward through bubbles and bubbles flow backward through wavefronts.

A **wavefront** is the part of a DATA or NULL wave propagation that is being stably maintained by a same phase request. A wavefront cannot be overwritten by a following opposite phase wavefront.

A **bubble** is the part of a DATA or NULL wave propagation that is no longer being stably maintained because the associated request is requesting the opposite phase wavefront. A bubble can and will be overwritten by a wavefront.

There are DATA wavefronts and DATA bubbles, NULL wavefronts and NULL bubbles. The signal relationships for each of the four cases are shown in Figure A.3. Transitions about to occur are indicated along the bottom of Figure A.3.

Logically Determined Design: Clockless System Design with NULL Convention Logic^{TM}, by Karl M. Fant
ISBN 0-471-68478-3   Copyright © 2005 John Wiley & Sons, Inc.

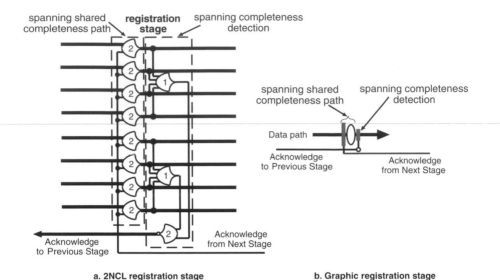

spanning shared completeness path

registration stage

spanning completeness detection

spanning shared completeness path

spanning completeness detection

Data path

Acknowledge to Previous Stage

Acknowledge from Next Stage

Acknowledge to Previous Stage

Acknowledge from Next Stage

a. 2NCL registration stage

b. Graphic registration stage

Figure A.1   2NCL registration stage and its symbolic representation.

A.3   WAVEFRONT PROPAGATION

Wavefront propagation occurs when DATA wavefronts spontaneously flow through (overwrite) NULL bubbles and when NULL wavefronts spontaneously flow through (overwrite) DATA bubbles. It is most convenient to view the wavefront flow and the bubble flow independently. Wavefronts flow along the data path through bubbles. Bubbles flow along the acknowledge path through wavefronts. Since a bubble is

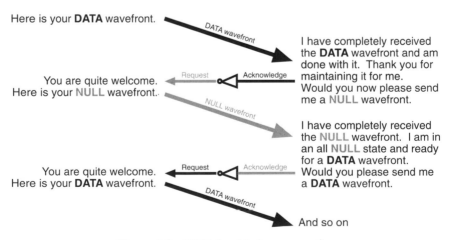

Here is your **DATA** wavefront.

DATA wavefront

I have completely received the **DATA** wavefront and am done with it. Thank you for maintaining it for me. Would you now please send me a **NULL** wavefront.

You are quite welcome. Here is your **NULL** wavefront.

Request   Acknowledge

NULL wavefront

I have completely received the **NULL** wavefront. I am in an all **NULL** state and ready for a **DATA** wavefront. Would you please send me a **DATA** wavefront.

You are quite welcome. Here is your **DATA** wavefront.

Request   Acknowledge

DATA wavefront

And so on

Figure A.2   2NCL handshake conversation.

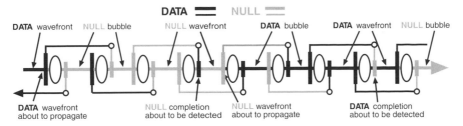

Figure A.3   Wavefronts and bubbles.

flowing along the acknowledge path, it alternately inverts its phase. A flowing bubble is alternately a DATA bubble then a NULL bubble, then a DATA bubble and so on. Figures A.4, A.5, and A.6 shows the details of wavefronts and bubbles flowing through two 2NCL registration stages.

A.4   EXTENDED SIMULATION OF WAVEFRONT FLOW

Figure A.7 is an extended simulation of wavefronts flowing through a four-stage pipeline that illustrates the mutual counterflow of wavefronts and bubbles. It begins with a DATA wavefront arriving at a pipeline with an extended NULL bubble. Sequences 1 to 10 illustrate wavefronts flowing through bubbles. The sequence beginning with 11 illustrates a bubble flowing through wavefronts.

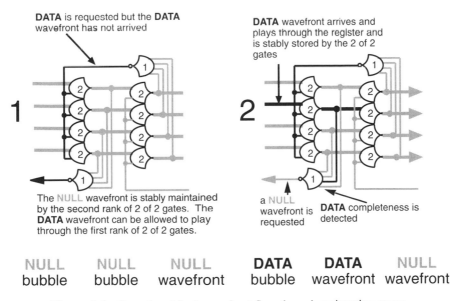

Figure A.4   Steps 1 and 2 of wavefront flow through registration stages.

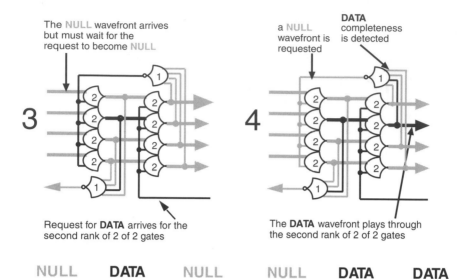

The NULL wavefront arrives but must wait for the request to become NULL

Request for **DATA** arrives for the second rank of 2 of 2 gates

a NULL wavefront is requested

DATA completeness is detected

The **DATA** wavefront plays through the second rank of 2 of 2 gates

NULL	**DATA**	NULL	NULL	**DATA**	**DATA**
wavefront	wavefront	bubble	wavefront	bubble	wavefront

Figure A.5   Steps 3 and 4 of wavefront flow through registration stages.

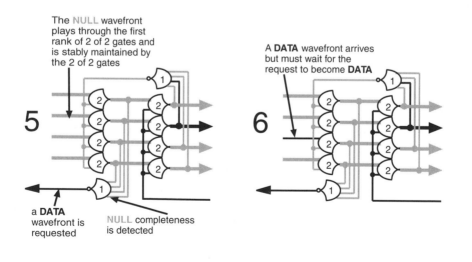

The NULL wavefront plays through the first rank of 2 of 2 gates and is stably maintained by the 2 of 2 gates

a **DATA** wavefront is requested

NULL completeness is detected

A **DATA** wavefront arrives but must wait for the request to become **DATA**

NULL	NULL	**DATA**	**DATA**	NULL	**DATA**
bubble	wavefront	wavefront	wavefront	wavefront	wavefront

Figure A.6   Steps 5 and 6 of wavefront flow through registration stages.

DATA ▬▬ **NULL** ▬▬

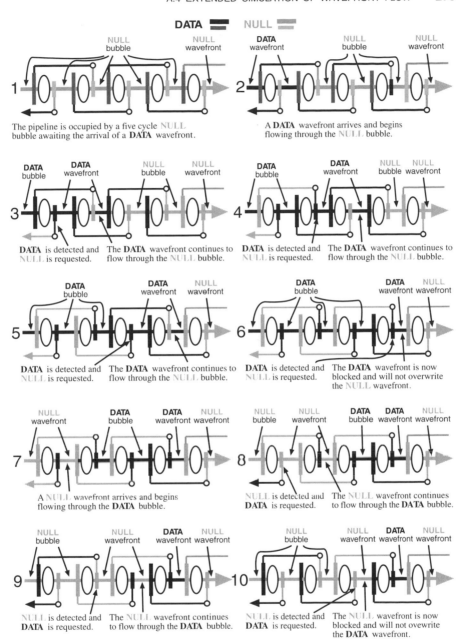

1 The pipeline is occupied by a five cycle NULL bubble awaiting the arrival of a **DATA** wavefront.

2 A **DATA** wavefront arrives and begins flowing through the NULL bubble.

3 **DATA** is detected and NULL is requested. The **DATA** wavefront continues to flow through the NULL bubble.

4 **DATA** is detected and NULL is requested. The **DATA** wavefront continues to flow through the NULL bubble.

5 **DATA** is detected and NULL is requested. The **DATA** wavefront continues to flow through the NULL bubble.

6 **DATA** is detected and NULL is requested. The **DATA** wavefront is now blocked and will not overwrite the NULL wavefront.

7 A NULL wavefront arrives and begins flowing through the **DATA** bubble.

8 NULL is detected and **DATA** is requested. The NULL wavefront continues to flow through the **DATA** bubble.

9 NULL is detected and **DATA** is requested. The NULL wavefront continues to flow through the **DATA** bubble.

10 NULL is detected and **DATA** is requested. The NULL wavefront is now blocked and will not ovewrite the **DATA** wavefront.

Figure A.7 Wavefront flow simulation.

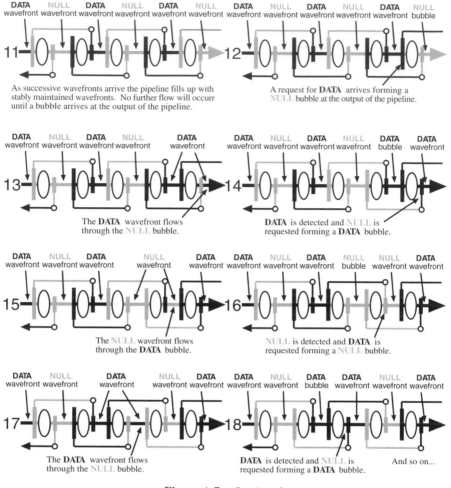

Figure A.7 *Continued.*

There are dynamic behaviors, in rings, for instance, where the behavior is most easily understood as bubbles flowing through wavefronts. This single bubble will continue flowing backward through the wavefronts alternating between a DATA bubble and NULL bubble. When it reaches the input of the pipeline, it will allow a new wavefront to enter the pipeline.

A.5 WAVEFRONT AND BUBBLE BEHAVIOR IN A SYSTEM

A system is generally initialized to an all NULL data path. This is a NULL bubble stretching completely through the system. The system is then ready to accept DATA wavefronts. DATA wavefronts can enter the system or be generated from within the system, from memory, for instance, and proceed flowing through the NULL bubbles. A convenient way to view system behavior is as DATA wavefronts flowing through a background of NULL bubbles. When wavefronts are not flowing through a pipeline, their quiescent waiting state is a NULL bubble.

It is possible to fill the system with DATA bubbles. If the input ceases cycling with a DATA wavefront stably presented to the input, the DATA wavefront will flow all the way through the system trailing DATA bubbles. When it gets to the output, the system data path will be in an all DATA state (full of data bubbles) awaiting the input of a NULL wavefront.

If the output ceases cycling, wavefronts will bunch up, filling the system with backed up wavefronts and no bubbles. Once full, the system will be inactive until the output begins cycling and bubbles begin flowing through the blocked wavefronts.

Playing with 2NCL

To use NCL for commercial purposes, a license must be obtained from Theseus Logic which includes macro libraries of operators that supports direct fabrication in recent CMOS technologies. An academic license is available for teaching institutions, and this includes libraries and tools. However, to get an initial understanding and experience of logically determined system design and behavior, anyone can quickly and conveniently begin running circuits in simulators and even mapping circuits into FPGAs.

The NCL operators can be represented in terms of SR flip-flops with Boolean set and reset functions, and logically determined systems can be reliably expressed in terms of these operators. While these representations are inefficient for commercial purposes, they are quite satisfactory for learning. They can be used with software circuit simulators and with FPGA implementations. Most of the NCL example circuits in this book were simulated using these macros in a simulation program called Circuit-maker [6].

The following examples provide the basics of mapping the NCL operators into the SR flip-flop representations. The SR flip-flop provides the state-holding behavior. A set pull-up or transition to DATA function transitions the flip-flop to DATA. The transition to DATA function is the Boolean implementation of the transition to DATA Boolean equation associated with the operator. A reset or pull-down or transition to NULL function transitions the flip-flop to NULL. The transition to NULL function is all NULL.

B.1 THE SR FLIP-FLOP IMPLEMENTATIONS

The SR filp-flop representation of N of N operators is shown in Figure B.1. These are identical to M of M C-elements. If the simulator has built in C-element representations then these may be used as the M of M operators.

The representation of 1 of N operators in shown Figure B.2. The 1 of N operator does not exhibit state-holding behavior and does not require the SR flip-flop.

Logically Determined Design: Clockless System Design with NULL Convention Logic^{TM}, by Karl M. Fant
ISBN 0-471-68478-3   Copyright © 2005 John Wiley & Sons, Inc.

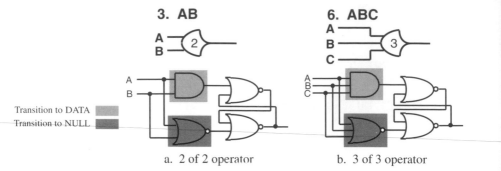

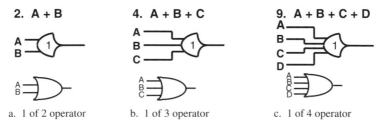

a. 2 of 2 operator   b. 3 of 3 operator

Figure B.1 *N* of *N* operators in terms of SR flip-flops.

The 1 of *N* operator is identical to the Boolean OR function, and Boolean OR functions can be used to directly represent them.

The SR flip-flop representation of *M* of *N* operators is shown in Figure B.3. The transition to DATA function is the Boolean implementation of the transition to DATA Boolean equation associated with the operator. The transition to NULL function for all operators is all NULL.

B.2 INITIALIZATION

Some operators must be initially forcible to DATA or to NULL to place a system in a known stable state. Initialization structures and protocols are described in Section 3.5 of Chapter 3. These initialization protocols depend on initializable operators.

The basic requirement is that when the init signal goes to DATA or HIGH, it forces one side of the SR flip-flop to LOW and the other to HIGH. When the init signal transitions to LOW, the operator becomes transparent to the init signal and behaves in relation to its transition to DATA and transition to NULL functions. Figure B.4 shows the general strategy for initializing operators represented in terms of SR flip-flops. The initialization can be integrated into the two functions. Figure B.5 shows examples of specific initialization functions.

a. 1 of 2 operator   b. 1 of 3 operator   c. 1 of 4 operator

Figure B.2 1 of *N* operators in terms of SR flip-flops.

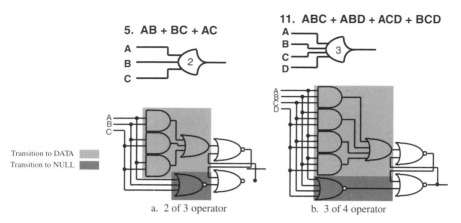

Figure B.3 *M* of *N* operators in terms of SR flip-flops.

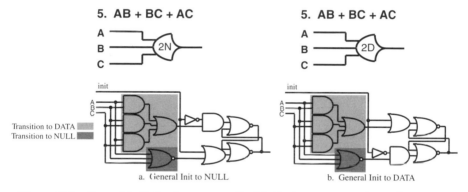

Figure B.4 General initialization strategy for SR flip-flop representations of operators.

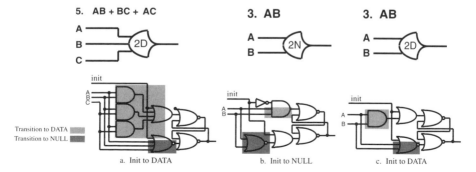

Figure B.5 Example of initializable operators.

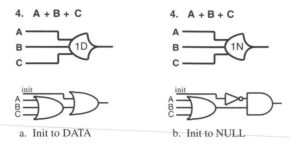

a. Init to DATA   b. Init to NULL

Figure B.6 Initializing 1 of N operators.

The 1 of N operators must also be forced to a specific state when, for instance, the initializing is by means of the acknowledge path. Figure B.6 shows the initialization for 1 of N operators.

B.3 AUTO-PRODUCE AND AUTO-CONSUME

Any NCL expression can be isolated and its inherent behavior exercised with auto-produce cycles on its inputs and auto-consume cycles on its outputs. If the auto-produce and auto-consume cycles have shorter periods than any cycle internal to the NCL expression, then they will not affect its behavior, and the inherent behavior of the expression will be exhibited. The auto cycles are presented in Section 13.2 of Chapter 13. The input acknowledge can also be used as a request to a data generator to produce varying input patterns.

Pipeline Simulation

The space-time diagrams used in Chapters 10 and 11 were generated by a Microsoft Excel [34] spreadsheet that simulates the behavior of the pipeline and produces space-time diagrams of the flow of successive wavefronts through successive pipeline cycles. The simulation models the forward propagation of data path wavefronts through the pipeline as well as the backward propagation of acknowledge signals through the pipeline. Although an acknowledge signal moves backward in the pipeline space, it progresses forward in time. So the signal relationships of multiple-pipeline stages can be arranged such that all signal relationships are progressive in time and thus expressible as a spreadsheet simulation. A pipeline ring, on the other hand, entails a circular reference and cannot be directly modeled by a spreadsheet simulation.

The delay for all component paths of the 13 cycles of the simulation can be individually set in a delay data table that drives the simulation, as shown in Figure C.1. Note that the vertical axis represents successive wavefronts entering the pipeline. The horizontal axis represents successive cycle components of the pipeline. The table details each delay associated for each wavefront at each pipeline registration stage. The column 'reg x' is the register delay. The column 'comp x' is completion

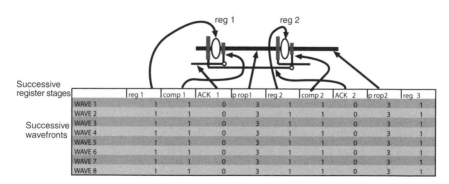

Successive register stages		reg 1	comp 1	ACK 1	prop1	reg 2	comp 2	ACK 2	prop2	reg 3
Successive wavefronts	WAVE 1	1	1	0	3	1	1	0	3	1
	WAVE 2	1	1	0	3	1	1	0	3	1
	WAVE 3	1	1	0	3	1	1	0	3	1
	WAVE 4	1	1	0	3	1	1	0	3	1
	WAVE 5	1	1	0	3	1	1	0	3	1
	WAVE 6	1	1	0	3	1	1	0	3	1
	WAVE 7	1	1	0	3	1	1	0	3	1
	WAVE 8	1	1	0	3	1	1	0	3	1

Figure C.1 Table of delays for spreadsheet simulation.

Logically Determined Design: Clockless System Design with NULL Convention Logic^{TM}, by Karl M. Fant
ISBN 0-471-68478-3   Copyright © 2005 John Wiley & Sons, Inc.

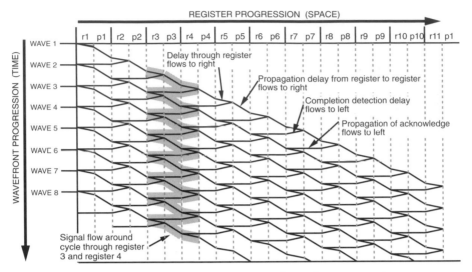

Figure C.2   Baseline space-time graph of wavefront propagation.

delay. The column 'ACK x' is the acknowledge delay. The column 'prop x' is the data path delay.

Figure C.2 is the baseline graph generated from the baseline delay table, showing the time-space relationships among the successive wavefronts propagating through the pipeline stages. A wavefront is represented by a line flowing from left diagonally down toward the right. The lines flowing from right to a lower left position are acknowledge signals. Columns in the graph represent the pipeline registration stages. Each registration stage is represented by two columns; rx is 'reg x' and px is 'prop x'. Wavefronts flow progressively forward through rx and px; r1, p1, r2, p2, r3, p3, ...; bubbles (acknowledge/request signals) flow backward beginning at the right edge of an rx column, which represents the completion delay 'comp x', and then flowing backward through $px - 1$ and $rx - 1$, which represents the acknowledge delay 'ACK x'. The left side of the graph is the input of the pipeline, and the right side of the graph is the output of the pipeline.

The Excel file can be obtained at *http://www.theseusresearch.com/ LDSD_Book_Interactive.htm.*

REFERENCES

1. K. van Berkel. Beware the isochronic fork. *Integration* 13 (1992): 103–128.

2. K. van Berkel and C. E. Molnar. Beware the three way arbiter. *IEEE J. Sol. State Cir.* 34 (June 1999): 840–848.

3. S. G. Browning. Tandem queues with blocking: A comparison between dependent and independent service. *Oper. Res.* 46 (May–June 1998): 424–429.

4. J. A. Brzozowski and Carl-Johan H. Seger. *Asynchronous Circuits.* New York: Springer-Verlag, 1995.

5. S. M. Burns. Performance analysis and optimization of asynchronous circuits. PhD dissertation. Cal Tech, Pasadena, 1991.

6. Circuitmaker, Protel Technology Inc., 5252 N. Edgewood Dr., Suite 175, Provo, Utah, 84604, *http://www.altium.com/circuitmaker/*.

7. J. Cortadella, M. Kishinevsky, A. Kondratyev, L. Lavagno, and A. Yakovlev. Petrify: A tool for manipulating concurrent specifications and synthesis of asynchronous controllers. *IEICE Trans. Info. Sys.* 80 (March 1997): 315–325.

8. D. L. Dill. *Trace Theory for Automatic Hierarchical Verification of Speed-Independent Circuits.* Cambridge: MIT Press, 1989, pp. 2, 3.

9. J. Ebergen and R. Berks. Response time properties of some asynchronous circuits. *Proc. IEEE* 87 (February 1999): 308–318.

10. K. M. Fant and S. A. Brandt. NULL Convention Logic™: A complete and consistent logic for asynchronous digital circuit synthesis. *International Conference on Application Specific Systems, Architectures and Processors.* IEEE Computer Society Press, Los Alamitos, CA, 1996, pp. 261–273.

11. K. Fant, A. Taubin, and J. McCardle. Design of delay insensitive three dimension pipeline array multiplier. *International Conference on Computer Design: VLSI in Computers and Processors.* 2002, pp. 104–111.

12. M. E. Fisher. Walks, walls, wetting and melting. *J. Stat. Phys.* 34 (May–June 1984): 667–729.

13. C. Foley. Characterizing metastability. *Second International Symposium on Advanced Research in Asynchronous Circuits and Systems.* IEEE Computer Society Press, Los Alamitos, CA, 1996, pp. 175–184.

14. P. J. Forrester. Exact results for vicious walker models of domain walls. *J. Phys. A. Math. Gen.* 24 (1991): 203–218.

Logically Determined Design: Clockless System Design with NULL Convention Logic™, by Karl M. Fant
ISBN 0-471-68478-3   Copyright © 2005 John Wiley & Sons, Inc.

15. J. D. Garside, S. B. Furber, and S.-H. Chung. Amulet3 revealed. *Fifth International Symposium on Advanced Research in Asynchronous Circuits and Systems.* IEEE Computer Society Press, Los Alamitos, CA, 1999, pp. 51–59.

16. M. R. Greenstreet and K. Steiglitz. Bubbles can make self-timed pipelines fast. *J. VLSI Signal Proc.* 2 (1990): 139–148.

17. M. Greenstreet and B. de Alwis. How to achieve worst case performance. *Seventh International Symposium on Advanced Research in Asynchronous Circuits and Systems.* IEEE Computer Society Press, Los Alamitos, CA, 2001, pp. 206–216.

18. R. W. Hall. Queueing methods: For services and manufacturing. Englewood Cliffs, NJ: Prentice Hall, 1991, ch. 10.

19. S. Harvard Computational Laboratory. *Synthesis of Electronic Computing Circuits.* Cambridge: Harvard University Press, 1951.

20. D. Helbing. Traffic and related self-driven many-particle systems. *Rev. Mod. Phy.* 73 (October 2001): 1067–1141.

21. S. P. Hoogendoorn and P. H. L. Bovy. Modeling multiple user-class traffic. *Transport. Res. Rec.* 1644, Paper 98-0692 (1998), pp. 57–69.

22. S. L. Hurst. *The Logical Processing of Digital Signals.* New York: Crane Russak, 1978.

23. D. Kearney and N. W. Bergmann. Performance evaluation of asynchronous logic pipelines with data dependant processing delay. In *Proc. 2nd Working Conf. on Asynch. Design Methodologies,* IEEE Computer Society Press, Los Alamitos, CA, May 1995, London, pp. 4–13.

24. M. Kishinevsky, A. Kondratyev, A. Taubin, and V. Varshavsky. *Concurrent Hardware: The Theory and Practice of Self-timed Design.* New York: Wiley, 1994.

25. A. Kondratyev, O. Roig, K. Fant, A. Taubin, and L. Neukom. Checking delay-insensitivity: 10K gates and beyond. In *Proc. Eighth Int. Symp. on Adv. Res. in Asynch. Circuits and Systems*, Manchester, April 2002, IEEE Computer Society Press, Los Alamitos, CA, pp. 149–157.

26. W. Kuang, J. S. Yuan, R. F. DeMara, D. Ferguson, and M. Hagedorn. A delay-insensitive FIR filter for DSP applications. In *Proc. Ninth An. NASA Symp. on VLSI Design*, Albuquerque, NM, November 8–9, IEEE Computer Society Press, Los Alamitos, CA, 2000, pp. 2.2.1–2.2.7.

27. P. M. Lewis II and C. L. Coates. *Threshold Logic.* New York: Wiley, 1967.

28. Michiel Ligthart, Karl Fant, Ross Smith, Alexander Taubin, and Alex Kondratyev. Asynchronous design using commercial HDL synthesis tools. In *Proc. Sixth Int. Symp. on Adv. Res. in Asynch. Circuits and Systems* (ASYNC 2000), Israel, April 2000, IEEE Computer Society Press, Los Alamitos, CA, pp. 114–125.

29. A. J. Martin. Programming in VLSI: From communicating processes to delay-insensitive circuits. In C. A. R. Hoare, ed., *Developments in Concurrency and Communication.* Reading, MA: Addison-Wesley, 1990, pp. 1–64.

30. A. J. Martin. A synthesis method for self-timed VLSI circuits. In *Proc. IEEE Int. Conf. on Computer Design*, Rye Brook, NY, October 5–8, 1987, IEEE Computer Society Press, Los Alamitos, CA, pp. 225–229.

31. A. J. Martin. The limitations to delay-insensitivity in asynchronous circuits. In *Proc. Sixth MIT Conf. on Adv. Res. in VLSI Processes*, MIT Press, Cambridge, MA, 1990 pp. 263–277.

32. T. H. Meng. *Synchronization Design for Digital Systems.* Boston: Kluwer Academic, 1991.

33. C. J. Meyers. *Asynchronous Circuit Design.* New York: Wiley, 2001.

34. *Microsoft Excel Users Guide.* Microsoft Corporation, Redmond, WA, 1994.

35. D. E. Muller. Asynchronous logics and an application to information processing. In H. Aiken and W.F. Main, eds., *Proc. Symp. on Applications of Switching Theory in Space Technology.* Stanford: Stanford University Press, 1963, pp. 289–297.

36. S. Muroga. *Threshold Logic and its Applications.* New York: Wiley-Interscience, 1971.

37. R. O. Onvural. Survey of closed queueing networks with blocking. *ACM Comput. Sur.* (June 1990): 83–121.

38. M. Pinedo and R. W. Wolff. A comparison between tandem queues with dependent and independent service times. *Oper. Res.* 30 (May–June 1982): 465–479.

39. C. V. Ramamoorthy and G. S. Ho. Performance evaluation of asynchronous concurrent systems using Petri nets. *IEEE Trans. Software Eng.* 6 (September 1980): 440–449.

40. C. L. Seitz. System timing. In Mead and L. Conway, eds., *Introduction to VLSI Systems.* Reading, MA: Addison-Wesley, 1980, pp. 242–262.

41. Semiconductor Industry Association. *International Technology Roadmap for Semiconductors.* San Jose, CA, 1999.

42. C. L. Sheng. *Threshold Logic.* New York: Academic Press, 1969.

43. R. Smith, K. Fant, D. Parker, R. Stephani, and C.-y. Wang. An asynchronous 2-D discrete cosine transform chip. In *Proc. Fourth Int. Symp. on Adv. Res. in Asynch. Circuits and Systems,* San Diego, April 1998, IEEE Computer Society Press, Los Angeles, CA, pp. 224–233.

44. R. Smith and M. Ligthart. Asynchronous logic design with commercial high level design tools. DAC, Los Angeles, CA, 2000.

45. S. C. Smith. Gate and throughput optimizations for NULL convention self-timed digital circuits. PhD dissertation. School of Electrical Engineering and Computer Science, University of Central Florida, 2001.

46. S.C. Smith. Speedup of self-timed digital systems using early completion. In *Proc. IEEE Comput. Soc. An. Symp. on VLSI,* April 2002, pp. 107–113.

47. S. C. Smith, R. F. DeMara, J. S. Yuan, M. Hagedorn, and D. Ferguson. Speedup of delay-insensitive digital systems using NULL cycle reduction. In *Tenth Int. Workshop on Logic and Synthesis,* June 2001, IEEE Computer Society Press, Los Alamitos, CA, pp. 185–189.

48. S. C. Smith, R. F. DeMara, J. S, Yuan, M. Hagedorn, and D. Ferguson. Delay-insensitive gate-level pipelining. *Integration* 30 (November 2001): 103–131.

49. S. C. Smith, R. F. DeMara, J. S. Yuan, M. Hagedorn, and D. Ferguson. NULL convention multiply and accumulate unit with conditional rounding, scaling, and saturation. *J. Sys. Arch.* 47 (February 2003): 977–998.

50. G. E. Sobelman and K. Fant. CMOS circuit design of threshold gates with hysteresis. In *Proc. Int. Symp. on Circuits and Systems,* June 1998, pp. 61–64.

51. L. Sorensen, A. Streich, and A. Kondratyev. Testing of asynchronous designs by 'inappropriate' means: Synchronous approach. In *Proc. Eighth Int. Symp. on Adv. Res. in Asynch. Circuits and Systems,* Manchester, April 2002, IEEE Computer Society Press, Los Alamitos, CA, pp. 171–180.

52. J. Sparsø and S. Furber, eds. *Principles of Asynchronous Circuit Design*. Boston: Kluwer, 2001.

53. J. Sparsø, J. Staunstrup, and M. Dantzer-Sørenson. Design of delay insensitive circuits using multi-ring structures. In *Proc. European Design Automation Conference*, 1992, Computer Society Press, Los Alamitos, CA. pp. 15–20.

54. K. Stevens, R. Ginosar, and S. Rotem. Relative timing. In *Fifth Int. Symp. on Adv. Res. in Asynchronous Circuits and Systems*, Barcelona 1999, IEEE Computer Society Press, Los Alamitos, CA, pp. 208–218.

55. I. E. Sutherland and S. Fairbanks. GasP: A minimal FIFO control. In *Proc. Seventh Int. Symp. on Adv. Res. in Asynch. Circuits and Systems*, Salt Lake City, 2001, IEEE Computer Society Press, Los Alamitos, CA, pp. 46–53.

56. I. E. Sutherland and J. K. Lexau. Designing fast asynchronous circuits. In *Proc. Seventh Int. Symp. on Adv. Res. in Asynch. Circuits and Systems*, Salt Lake City 2001, IEEE Computer Society Press, Los Alamitos, CA, pp. 184–193.

57. I. E. Sutherland. Micropipelines. *Comm. ACM* Vol. 32 (June 1989): 720–738.

58. S. H. Unger. *Asynchronous Sequential Switching Circuits*. New York: Wiley-Interscience, 1969.

59. V. I. Varshavsky. *Self-Timed Control of Concurrent Processes*. Dordrecht, The Netherlands: Kluwer Academic, 1990.

60. T. Verhoeff. Delay insensitive codes—An overview. *Distribu. Comput.* 3 (1998): 1–8.

61. T. Williams. Self-timed rings and their application to division. PhD dissertation. Department of Electrical Engineering and Computer Science, Stanford University, 1991.

62. I. Ziedins. Tandem queues with correlated service times and finite capacity. *Math. Oper. Res.* 18 (November 1993): 901–915.

Logically Determined Design: Clockless System Design with NULL Convention Logic^{TM}, by Karl M. Fant
ISBN 0-471-68478-3   Copyright © 2005 John Wiley & Sons, Inc.